Entrepreneurial Ecosystems in Unexpected Places

Veneta Andonova • Milena S. Nikolova
Dilyan Dimitrov

Entrepreneurial Ecosystems in Unexpected Places

Examining the Success Factors of Regional Entrepreneurship

palgrave
macmillan

Veneta Andonova
School of Management
Universidad de Los Andes
Bogotá, Colombia

Milena S. Nikolova
American University in Bulgaria
Blagoevgrad, Bulgaria

Dilyan Dimitrov
Sofia, Bulgaria

ISBN 978-3-319-98218-2 ISBN 978-3-319-98219-9 (eBook)
https://doi.org/10.1007/978-3-319-98219-9

Library of Congress Control Number: 2018952373

This Palgrave Macmillan imprint is published by the registered company Springer Nature Switzerland AG
The registered company address is: Gewerbestrasse 11, 6330 Cham, Switzerland

Acknowledgements

This book would not have been possible without the support, involvement and contribution of many different individuals and organizations.

The first inspiration for this book came from Bill Aulet, Managing Director of the Martin Trust Center for MIT Entrepreneurship and Professor of the Practice at the MIT Sloan School of Management. In a conversation about the ways to accelerate regional entrepreneurial processes, Bill talked to us about the importance of celebrating local success stories and using them to inspire the startup community, as well society at large. This conversation was the start of our journey in writing this book.

We wish to thank Jakob Modéer and the entire regional team of the Swiss Entrepreneurship Program for their generous support, which made it possible for us to meet and talk to different members of the entrepreneurial community across the Western Balkans. We also appreciate the help provided by the Bulgarian Private Equity and Venture Capital Association (BVCA) and specifically its Director, Maria Marinova, who brokered valuable introductions in the investor community and who helped us conduct many interviews during the data collection phase.

We derived many insights from the detailed data provided by Eleven Ventures and LAUNCHub—the two accelerators that played a catalysing role for startup dynamics throughout the region, serving as springboard for many ventures across Eastern Europe. Without this information, much of the analysis that is specific to this region and that is woven throughout the book would have been impossible.

We received invaluable support from a team of research assistants who worked on this project. Valya Nadova, Desislava Papazova, Kristina Panayotova, Eloisa Isaza Rodriguez, Camilo De Los Rios, Jonathan Pérez, Daniel Serpa, Maria Mora and Alejandro Arias carried out a lot of the data processing, statistical analyses and the extraction of insights. Georgi Chapanov supported the carrying out of some of the interviews.

This was a huge learning journey for us. The interviews we conducted with different members of the ecosystems across countries in the region revealed the depth and richness of the processes that have led to what we are able to see today. We are deeply grateful to the many entrepreneurs, investors, startup experts, public officials and researchers who made time to talk to us and who shared their memories, opinions and dreams. Entrepreneurship is all about the people, and without the energy, smarts and boldness they have demonstrated over the years, we would not have had a story to tell today. We thank (in alphabetical order) Alema Pelesic, Alen Sage, Andi Gjoutaj, Andrea Mirnic, Andrej Hanzir, Bogdan Iordache, Bogdan Mijovic, Bruno Omeromic, Dan Lupu, Daniel Lorer, Daniel Tomov, Dejan Aleksov, Dragana Neshovska, Dritan Mezini, Eldin Alimanovic, Elena Hadzi Decova, Elena Nikolova, Elvin Guri, Emir Slezovic, Enxhi Dyrmishi, Ertsan Cabiri, Evgeny Angelov, Florin Talpeş, George Petrov, George Roth, Georgi Ivanov, Georgi Kadrev, Georgi Monolov, Gerti Boshnjaku, Grigoris Chatzikostas, Iskren Krusteff, Ivan Brkan, Ivan Gligorijevic, Ivana Stankovic, Ivaylo Simov, Ivo Spigel, Jakob Modéer, Jan Zlatan Kulenovic, Jonel Kristo, Joniada Hito, Kiril Petkov, Konstantin Hristov, Konstantin Nikolov, Konstantin Yordanov, Kosta Andric, Kresimir Kukec, Larry Biehl, Lilly Drumeva, Luka Prisunjak, Lyubomir Yanchev, Marius Ursache, Marko Misulic, Max Gurvits, Michael Bohanes, Milena Stoycheva, Milos Dosen, Milos Lazovic, Miloš Milisavljević, Mira Krusteff, Mircea Pasoi, Mitko Cabevski, Momchil Vasilev, Nancy Schiller, Natasha Petrova, Nebojsa Lazic, Nikola Yanev, Nikolay Andonov, Nina Angelovska, Nina Nikolikj, Orlin Radev, Pavel Ezekiev, Petar Ninovski, Petar Sharkov, Petko Ruskov, Rossen Plevneliev, Samira Nuhanovic, Shkenlzen Marku, Svetoslav Pondev, Svetozar

Georgiev, Svilen Rangelov, Teddy Pejoski, Todor Breshkov, Todor Lilkov, Vesna Agic, Viktor Dano, Vladimir Rusev and Volen Vulkov.

Veneta and Milena would like to thank the many wonderful students of the American University in Bulgaria for being a powerful source of inspiration. Their entrepreneurial creativity, energy and passion fuelled our desire to write this book as a contribution to the growing ecosystem that attracts these young talents and encourages them to continue to put the region on the global entrepreneurship map.

Last but not least, we would like to thank Palgrave Macmillan for placing their trust in us and for helping us to tell the story of the young entrepreneurial ecosystem that is emerging in the Balkans, which will hopefully inspire other unlikely places to unlock their startup potential. We extend special thanks to Lucy Kidwell, Senior Editorial Assistant, whose support and guidance were essential in making this book a reality.

25 June 2018

Veneta Andonova
Milena S. Nikolova
Dilyan Dimitrov

Contents

Part I Entrepreneurial Ecosystems in Numbers and Quotes 1

1 What Is an Entrepreneurial Ecosystem? 3
Introduction 3
What Is an Entrepreneurial Ecosystem? 4
How Can the Construct of Entrepreneurial Ecosystem Be so Clear and Empowering to Insiders but so Elusive and Vague to Outsiders? 5
Why Entrepreneurial Ecosystems in the Balkans? 7
So What's New? 9
Data and Research Method 11
What Did We Learn in a Nutshell? 12
References 15

2 Entrepreneurial Ecosystems in the Balkan Peninsula 17
Introduction 17
Why Would a Chinese Peroxide Company Pay US$1 Billion for a Slovenian Talking Feline? 18
Key Entrepreneurship Indicators 19
The Entrepreneurial Framework Conditions of the GEM 21
Government Programmes and the Public Sector 23

Infrastructure, Human Capital and Innovation Capabilities 27
Universities and Entrepreneurship in the Balkans 29
Talent Is Abundant 30
Entrepreneurial Finance 33
Eleven, LAUNCHub and the Accelerator Experience 35
Looking for Regional VC Investors 38
Market Functioning and Private Sector Sophistication 39
Startup Internationalization in the Balkans 39
Multinationals and the Labour Market 41
Culture 43
Media 45
Academic and Research Institutions and R&D Transfer 46
Technological Entrepreneurship 47
The Entrepreneurial Ecosystem in Romania 50
Looking to the Future 51
References 52

3 The Creation and Development of the Ecosystem: The Role of ICT 57
Introduction 57
Current Entrepreneurship Dynamics and Key Players in the Region 58
Historical Background 61
The (Re)Birth of Entrepreneurship on the Balkans 63
The Growth of the IT Industry 67
The Development of a Young Ecosystem 71
Entrepreneurial Finance 72
Finance for the Full Growth Cycle 73
Local or Foreign Investment Sources 74
Government Programmes and the Public Sector 75
Weaknesses and Challenges with Governments as Ecosystem Members 78
Academic and Research Institutions (and R&D Transfer) 78
The Lack of 'What Is Next' Thinking 79
Entrepreneurial Education at Early Stages 80
Marketing and Business Culture 81

Market Functioning and Private Sector Sophistication 82
Limited Market Size 86
Infrastructure 88
Culture 93
The Role of Media 98
References 99

4 **Balkan Startups and the Entrepreneurs Behind Them** 101
Introduction 101
Balkan Entrepreneurial Ecosystems: A First Approximation 104
Balkan Ventures in the Entrepreneurship Database Program at Emory University 106
Baltic Ventures 112
Central European Ventures 114
Startup Genome Report 117
The Startup Genome of Balkan Ventures 120
Developing Strategic Thinking in Entrepreneurs in the Balkans 133
References 135

5 **Founder and Stakeholder Interests: Combining Economic and Societal Factors in the Balkan Peninsula** 137
Introduction 137
The Genesis of Early-Stage Ventures 139
Entrepreneuring Profit and Purpose 139
Business Models and Business Model Innovation 142
Values and Entrepreneurship: Startup Founders' Identities and Investors' Interests 145
Founders' Identities and Sensemaking 146
Market Stakeholder Interests 148
Context, Data and Methods 149
Case Study Analysis 154
Stage 1: Experiences (Personal and Proximate) 155
Stage 2: Purpose 158
Stage 3: Team Formation 160

Stage 4: Interaction with Market Stakeholders 162
Stage 5: Balancing Purpose and Return in Business Model Innovation 165
Conclusion 169
References 173

Part II Case Studies 183

6 Coherent Labs 191
Company Profile 191
Case Study 192
Lessons Learned 193
Founder Interview 193
Investor Interview 202

7 Dronamics 211
Company Profile 211
Case Study 212
Lessons Learned 214
Founder Interview 214
Investor Interview 224

8 Enhancv 231
Company Profile 231
Case Study 232
Lessons Learned 233
Founder Interview 234
Investor Interview 246

9 FITE TV 249
Company Profile 249
Case Study 250
Lessons Learned 251
Founder Interview 251
Investor Interview 261

10 Sensika Technologies 265
Company Profile 265
Case Study 266
Lessons Learned 268
Founder Interview 269
Investor Interview 278

11 Rentl.io 285
Company Profile 285
Case Study 286
Lessons Learned 288
Founder Interview 288
Advisor Interview 295

12 Telerik 301
Company Profile 301
Case Study 302
Founder Interview 302

Part III Conclusions 311

13 Concluding Thoughts and Lessons Learned from the Birth of Entrepreneurial Ecosystems in Unlikely Places 313
Why Did We Write This Book? 313
Our Own Story and Lessons Learned 315
What Mattered? 317
What Did Not Matter (or Mattered Less)? 320
Entrepreneurship as Spirit and Way of Thinking 322
Ambitions and Being Yourself 325

Index 327

List of Figures

Fig. 1.1	Entrepreneurial ecosystem at a glance. Source: Adapted from Isenberg (2011), Brown and Mason (2017) and GEM Bulgaria 2015/16 Country Report	7
Fig. 3.1	The entrepreneur in the Balkan region	59
Fig. 4.1	For how long have team members known each other? Source: Startup Genome Survey Data, 2014, 56 startups Eleven Accelerator Venture Fund	123
Fig. 4.2	Percentage of technical team members out of total team size. Source: 56 startups of the portfolio of Eleven	124
Fig. 4.3	Focus of the team. Source: 56 startups of the portfolio of Eleven	125
Fig. 4.4	Startups' current focus. Source: Startup Genome Survey Data, 2014, 56 startups Eleven Accelerator Venture Fund	125
Fig. 4.5	Key challenges startup is facing today. Source: Startup Genome Survey Data, 2014, 56 startups Eleven Accelerator Venture Fund	126
Fig. 4.7	Primary geographical market for product/service. Source: Startup Genome Survey Data, 2014, 56 startups Eleven Accelerator Venture Fund	128
Fig. 4.8	Next geographical region considered for international expansion. Source: Startup Genome Survey Data, 2014, 56 startups Eleven Accelerator Venture Fund	128

Fig. 4.9 How they plan to become active in the next geographical region. Source: Startup Genome Survey Data, 2014, 56 startups Eleven Accelerator Venture Fund 129
Fig. 4.10 Time expected to wait until they move to next geographical region. Source: Startup Genome Survey Data, 2014, 56 startups Eleven Accelerator Venture Fund 129
Fig. 4.11 Completion of development phases during Discovery stage. Source: Startup Genome Survey Data, 2014, 56 startups Eleven Accelerator Venture Fund 130
Fig. 4.12 Completion of development phases during Validation stage. Source: Startup Genome Survey Data, 2014, 56 startups Eleven Accelerator Venture Fund 131
Fig. 4.13 Completion of development phases during Efficiency stage. Source: Startup Genome Survey Data, 2014, 56 startups Eleven Accelerator Venture Fund 132
Fig. 4.14 Completion of development phases during Scale stage. Source: Startup Genome Survey Data, 2014, 56 startups Eleven Accelerator Venture Fund 132
Fig. 4.15 Consistent vs. inconsistent startups. Source: Startup Genome Survey Data, 2014, 56 startups Eleven Accelerator Venture Fund 133
Fig. 4.16 Part-time employees in consistent and inconsistent startups. Source: Startup Genome Survey Data, 2014, 56 startups Eleven Accelerator Venture Fund 133
Fig. 13.1 Visual summary of the factors identified by regional ecosystem leaders as contributing to the growth of entrepreneurship in the region 317
Fig. 13.2 Factors identified by regional ecosystem leaders as influencing entrepreneurship in the region 324

List of Tables

Table 2.1	Total entrepreneurship activity reported in the Global Entrepreneurship Monitor Report of 2017	20
Table 2.2	Entrepreneurial framework condition scores as reported in the National Experts Survey of GEM, weighted average (1 = highly insufficient, 5 = highly sufficient)	22
Table 2.3	Average expert ratings for government policies as reported in the National Experts Survey, weighted average (1 = completely false, 5 = completely true)	24
Table 2.4	Perceived support by government and quality of policies reported in the Global Entrepreneurship Monitor Report of 2017	25
Table 2.5	World Bank Doing Business 2017—Starting a Business (average results for sample groups)	25
Table 2.6	Global Innovation Index 2017—Infrastructure Pillar (average score and ranking for sample groups)	27
Table 2.7	Average yearly wages, salaries and benefits paid to workers in US dollars (2013–17, US$)	32
Table 2.8	Highest level of education completed (in years) by the founder with the most education (2013–17)	33
Table 2.9	Highest level of education completed by the founder with the most education (2013–17)	33
Table 2.10	Relative importance of accelerator benefits as perceived by entrepreneurs—Balkan states (2013–17)	36

Table 2.11	Relative importance of accelerator benefits as perceived by entrepreneurs—Baltic states (2013–17)	37
Table 3.1	Economies in the Balkans (in US$)	86
Table 4.1	Stated headquarters and the origin of the founders	107
Table 4.2	Legal status of startup ventures	107
Table 4.3	What is the current operating model of your venture?	108
Table 4.4	Which primary sector is being impacted by your venture's activities?	109
Table 4.5	Industries reported by LAUNCHub applicants	110
Table 4.6	Which of the following impact objectives does your venture currently seek to address?	111
Table 4.7	Anticipated profit margin for firms	112
Table 4.8	Anticipated financial goals for firms	112
Table 4.9	Average expected investment of firms by source in US dollars	113
Table 4.10	Stated headquarters of Eleven application sample	121
Table 4.11	Stated headquarters of accepted ventures	123
Table 5.1	Startups with a recognizable social and/or environmental purpose	151
Table 5.2	Categories, frequency and meaning	155
Table 5.3	Timeline of second-order categories leading to business model innovation	156

Part I

Entrepreneurial Ecosystems in Numbers and Quotes

1

What Is an Entrepreneurial Ecosystem?

Introduction

This book tells the story of the young Balkan startup ecosystem and the mindset change that was provoked by its pioneers. It celebrates entrepreneurship by showing how it can be embraced as a value and a way of thinking in societies with limited entrepreneurial traditions. The text offers insights and an analytical overview of the processes that occurred after 2000 and the factors that enabled Bulgaria to become the startup capital of the Balkan region. It also outlines some of the factors that will be critical to sustaining this success in the future in the complex global and European context.

One of the key contributions of the book is to identify the achievements of the regional ecosystem that go beyond the financial numbers: this perspective is essential to an understanding of the profound change brought to countries with a newly discovered entrepreneurial spirit. In this, the book uncovers the role that the leading accelerator venture funds played in the creation of a vibrant regional community of entrepreneurs. The book also features remarkable real-life case studies and insights from

V. Andonova et al., *Entrepreneurial Ecosystems in Unexpected Places*,
https://doi.org/10.1007/978-3-319-98219-9_1

many of those involved in regional startups as well as programmes and investment funds in the region.

In telling the story of how the young entrepreneurial ecosystem in the Balkans started and has continued to grow, this book offers lessons learned that will be invaluable to would-be entrepreneurs and decision makers in other countries which are now on the path of designing and nurturing entrepreneurial ecosystems. It also provides insights into the relevant measures of success that capture benefits beyond the usual monetary indicators, which are typically applied in more mature ecosystems but which are not as meaningful in younger ecosystems, where importance should be placed on celebrating a new entrepreneurial mindset and changing values. It also addresses the need of existing ecosystem participants to better understand their role and to prepare them for the future.

What Is an Entrepreneurial Ecosystem?

The construct of entrepreneurial ecosystems has grown in popularity, appealing to the imaginations of consultants, policy-makers, educators, investors, influencers, researchers, opinion leaders and entrepreneurs alike. However, it is probably only academic researchers who would seek to clarify the construct before answering questions or sharing their views about a specific entrepreneurial ecosystem or about this phenomenon more generally. Therefore, a very intriguing situation arises: on the one hand, most people identifying themselves as participants in an entrepreneurial ecosystem would feel confident when talking about it, while most rigorously trained academic outsiders, on the other hand, would refer to the concept as a 'fad', a 'chaotic' or 'fuzzy' construct, or dismiss it altogether, arguing that it lacks conceptual limitations, rigour or differentiation from other already well-established and equally popular concepts (Martin, 2015). The middle ground is taken by those who believe that entrepreneurial ecosystems are 'highly variegated, multi-actor and multi-scalar', so that the best scholarly researchers can do is to describe different experiences because comparing, extrapolating and transposing policies and experiences would not be pertinent (Brown & Mason, 2017).

The lack of a commonly accepted definition of the term 'entrepreneurial ecosystem' has not been a problem for practitioner communities, whilst policy reports using the term have also proliferated. It would appear that the characteristics of entrepreneurial ecosystems in these communities typically include various actors, processes and institutions, which are blended together and put into a specific cultural, geographical and time context.

Moreover, practitioners and policy influencers have not only agreed that the construct of the entrepreneurial ecosystem is a useful one, but they have also developed insights and best practice to enable and effectively nurture them. Ideas such as 'Think big, start small, move fast' and 'Avoid artificially segmenting the community of entrepreneurs' have been advanced as practical avenues for building vibrancy in entrepreneurial ecosystems (Auerswald, 2015).

How Can the Construct of Entrepreneurial Ecosystem Be so Clear and Empowering to Insiders but so Elusive and Vague to Outsiders?

Part of the answer to this question lies in the sheer complexity of the concept and the ability of different stakeholders to deal with it. In entrepreneurial ecosystems, the entrepreneurs are the central actors, with a leading structuring role in entrepreneurial activity, which can really only be understood from a holistic and systemic perspective (Borissenko & Boschma, 2016), in that different layers of contexts interact and shape entrepreneurship. This is because it takes place in a community of interdependent actors (Stam, 2015) who belong to the broader cultural environment within which entrepreneurial expectations are shaped (Schmutzler, Andonova, & Díaz-Serrano, 2018) and entrepreneurial ventures operate (Mason & Brown, 2014). The social, institutional and relational aspects of the limited location in which a particular entrepreneurial ecosystem operates are the most significant explanatory variables for understanding the degree of entrepreneurial activity both for academic researchers and practitioners. Moreover, most of them will agree that the antecedents of the entrepreneurial ecosystems are deeply embedded in past historical and institutional choices and therefore are markedly path dependent (Feldman & Braunerhjelm, 2007).

The direction we take in this book is inspired in part by the practitioner approach to the concept of entrepreneurial ecosystems. We are attracted to the type of systemic thinking that describes how actors, institutions and processes blend in unique combinations to create multiple equilibria and beautiful chaos, which is fertile ground for business experimentation and triggers inspiring success stories. At the same time, we complement this enthusiasm and drive for story-telling with a well-structured, data-driven research strategy that takes a measured and scientific approach to this domain, thereby conceptually enriching the concept of entrepreneurial ecosystems. To this end, we try, as far as possible, to structure the available information into frameworks that will be familiar to scholarly oriented observers.

We analyse an entrepreneurial ecosystem in the light of the interactions that take place between six domains and their pertaining actors: (i) *entrepreneurial finance*—never the most important ingredient of an entrepreneurial ecosystem, but it is a necessary contributor to experimentation, scaling and ultimately positive economic outcomes and exits that make headlines; (ii) both *government programmes and agencies* support and shelter the entrepreneurial spirit against volatile market trends and fashions, giving stability and 'flow' to the system; (iii) *academic and research institutions* as well as R&D transfer capabilities ensure that a healthy flow of marketable innovations arrive at the product–market fit phase, from where scaling up is possible; (iv) market-supporting institutions and the degree of sophistication of the competitive strategies and business processes in the *private sector* represent a measurement of how fertile the local soil is for the germination of a grassroots success story locally. The strategic thinking of business leaders seems to be complementary to the entrepreneurial insights that emerge in entrepreneurial ecosystems (Zahra & Nambisan, 2012). This, together with (v) the *infrastructure*—physical, legal and human—represents the most tangible elements that nurture entrepreneurial activities. However, it is (vi) the *entrepreneurial mindset shaped by local specificities* that brings to life and shapes the new venture creation process. It is as much the proximate environment, comprised of family members and role models, as the distal cultural environment that reflects social norms and values that matter for the shaping of entrepreneurial intentions and ultimately entrepreneurial activity (Schmutzler et al., 2018) (see Fig. 1.1).

Why Entrepreneurial Ecosystems in the Balkans?

In the words of Winston Churchill, 'The Balkans produce more history than they can consume'. The intention of the author of this frequently cited quote was to highlight the disruptive and occasionally destructive

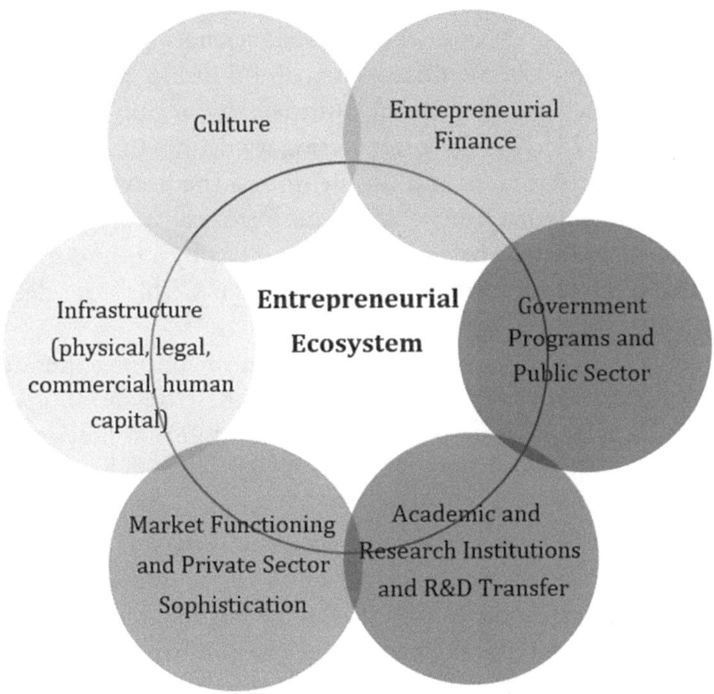

Fig. 1.1 Entrepreneurial ecosystem at a glance. Source: Adapted from Isenberg (2011), Brown and Mason (2017) and GEM Bulgaria 2015/16 Country Report

power of this part of the continent in history. More recently, at the EU-Western Balkans Summit in Sofia in 2018, Donald Tusk, the president of the European Council, the principal representative of the European Union, referred to the area of the Western Balkans in relation to its possible membership of the European Union in the following way: 'The region's economy and number of citizens cannot be a problem on the road to the EU but the Western Balkans has more problems per capita than Germany and France combined.' In the last 100 years, this part of Europe has frequently been seen as a nuisance rather than an asset to the wellbeing and prosperity of the Western part of the continent. In fact, many of the countries on the Balkan peninsula were at the military and ideological forefront of the Cold War. The case of the false accusations against Bulgaria in the attempted assassination of Pope John Paul II was amongst the most widely covered examples of how critical the Balkans were in the Cold War years. After the collapse of the Communist regimes in 1989, new and divisive lines appeared, separating the EU member countries from non-EU members. The painful disintegration of the former Yugoslavia led to wars and re-affirmed this region of Europe as a messy place. The Merriam-Webster dictionary defines the word *balkanize* as 'to break up into smaller and hostile units'. The heavy burden of pessimism and gloom that comes with the historical protagonism of the Balkans in world history has been recently offset by a beam of optimism coming from the local entrepreneurial ecosystems. This story deserves to be told.

The concept of an ecosystem is used within the business literature as a metaphor, communicating the idea that business organizations can be seen not only as pertaining to a single industry but that they can be perceived as interconnected elements of a 'larger system that spans many industries' (Neumeyer & Corbett, 2017). Arguably, the value of using this metaphor is to communicate to policy-makers and business people in general, including entrepreneurs, that it is the coordination and collective efforts of various stakeholders that make possible the success of individual endeavours and the increase of societal wellbeing that comes with it. As such, the metaphor vividly illustrates the need for 'networks and connections between individuals, organizations, and institutions as they cooperate on entrepreneurial ambitions' (Neumeyer & Corbett, 2017).

During the second half of the twentieth century, the Balkan countries were ruled by a particular style of collectivistic ideology that gave rise to the so-called Communist regimes, a type of dictatorship that relied on planned economies. These regimes were instituted after the Second World War and entailed the imprisonment of economic and intellectual elites as well as the collectivization of private property. The fall of the Communist regimes in the 1990s saw the creation of institutions for the protection of private property and, in many cases, the restitution of confiscated assets. The ideology of collectivity was rapidly substituted with the idea of the primacy of individual rights which, in many cases, appeared as an irrational backlash to the previous social order. Nevertheless, economic and social progress requires a certain, and frequently high, level of collectivity and an agreed-upon social purpose. We see the current rise of the entrepreneurial ecosystem in the Balkans as a restoration and legitimation of the need for collective and coordinated economic and social interactions, but on a different basis, that of individual freedom and entrepreneurial ambition. As such, the Balkans represent a laboratory for understanding how entrepreneurial ecosystems can emerge from hostile and sometimes violent historical periods, as they represent a metaphor for the creation of a collectively optimized infrastructure, institutions and social capital.

So What's New?

The entrepreneurial ecosystems approach seeks to understand the environment around the entrepreneurs and entrepreneurship in an economy and appreciate its effects on performance (Acs et al., 2017) but also on medium- and long-term structural forces that might be even more important. A key element is the interdependence of actors and factors in a particular community that leads to value creation in a certain territory, that is, entrepreneurship is the output of the entrepreneurial ecosystem. The entrepreneurial ecosystem approach brings to the forefront a mode of self-organization within the entrepreneurial ecosystem that is based on shared community values and norms and 'servant' leadership behaviour by community/publicly oriented entrepreneurs with a long-term commitment to a region (Acs et al., 2017). As such, entrepreneurship can be seen not only the result of the system but also an ingredient of a virtuous circle that leads to even more intense entrepreneurial activity (Mason & Brown, 2014; Stam, 2015).

In this instance, it is useful to apply the broader view of entrepreneurial ecosystems as business ecosystems that are particularly prone to support new venture creation. Here, the emphasis is not so much on new venture creation but on the interaction between well-established incumbents and new ventures which, in turn, may be corporate-sponsored or launched by independent entrepreneurs (Zahra & Nambisan, 2012). According to this more inclusive approach, incumbents and new ventures take different but often complementary roles within the business ecosystem and, as a result, exploit different opportunities that arise in the ecosystem (Livingston, 2007). Consequently, there is an increase in the diversity of business activity in the ecosystem, which adds dynamism and vibrancy and places it on a more opportunity-rich evolutionary path. This desirable outcome occurs only if all participants in the business ecosystem engage in thinking strategically about it and intentionally develop relationships with each other, in a process that balances individual firm opportunities and the development of the ecosystem as a whole. This balancing act comes as a result of strategic thinking, a capability that business ecosystems participants ideally hold.

Strategic thinking does not come easily or naturally. It is a process that transcends time, space and resources to offer innovative solutions to hard problems, and it has a number of qualities or elements, including entrepreneurial attitude and creativity, long-term orientation, a systemic view about problem solving, scenario planning and intention and direction, which involve visualizing the future before it happens. Thinking strategically about the business ecosystem and the role of entrepreneurship within it is instrumental for understanding the future and role of new ventures in the regional and global landscape. In this regard, the Balkans are one of the most fascinating places for exploring how entrepreneurial ecosystems are born and evolve. Ancient and recent historical events, such as wars and enduring diplomatic animosities, have been particularly damaging for the economic and social advancement of these nations and the region as a whole. Against this backdrop, the development of young entrepreneurial ecosystems offers the promise of the emergence of a new elite with the much-needed entrepreneurial mindset and strategic thinking that the region has traditionally lacked.

Data and Research Method

The research undertaken in the preparation of this book draws on a number of data sources and features to differing extents the following countries: Albania, Bosnia Herzegovina, Bulgaria, Croatia, Greece, Kosovo, Macedonia, Rumania, Serbia and Slovenia.

The Global Entrepreneurship Monitor (GEM) Consortium is a worldwide study on entrepreneurship that was first conceptualized in 1997. GEM does not just look at businesses but also at individuals between the ages of 18 and 64 years from a demographically representative portion of the population uncovering their attributes, aspirations, attitudes, perceptions and intentions.

The Entrepreneurship Database Program at Emory University contains a sample of 8666 entrepreneurs who applied to 82 accelerators and support programmes, monitored since 2013.

Eleven is an accelerator venture fund. Eleven was approved as fund manager of the JEREMIE (Joint European Resources for Micro to Medium Enterprises), a European Investment fund's accelerator and seed fund instrument in early 2012. Eleven held 10 application windows during its investment period, which ended in 2016, receiving about 3500 applications and approving 116 investments, mostly from Southeastern Europe. For its three years of operations, Eleven invested almost €12 million and has become one of the top three European accelerators in both the number of investments and the amounts invested (Eleven quarterly report, 2017).

LAUNCHub started as an accelerator venture fund in 2012 with a budget of €9 million. It had six team managing partners who came from different backgrounds and had related investment experience in consulting, finance, entrepreneurship and general management. As of 2017, LAUNCHub had a total portfolio value of €92 million. They evaluated 3000 investment opportunities of which 63 were chosen (2% acceptance rate); of these, 30 were active in 2018. They had had two exits and 31 write-offs before 2018.

More than 70 face-to-face interviews gave us the opportunity to gain deep insights and uncover the thinking processes of key stakeholders in the Balkan entrepreneurial ecosystems. These interviews turned out to be a reflection of personal passions, societal missions, team and community building and, much to our surprise, an extraordinary level of optimism about the region. In the following chapters, we share some of the most memorable parts of many of these conversations.

What Did We Learn in a Nutshell?

Entrepreneurial ecosystems cannot be properly measured by traditional economic performance variables for a number of reasons. First, in the case of ecosystems that are at an early stage of growth, economic and social value is created long before traditional measures are able to capture the impact of productivity or a growth in the number of startups. New ventures that populate these entrepreneurial ecosystems, and interact with multiple actors in the broader business and social spheres, have a subtle, intangible effect on cognitive predispositions such as the entrepreneurial intentions of others and subsequently on their entrepreneurial behaviour. Their early successes often constitute them not only as angel investors, a role that can be measured, but into role models, influencers and networkers, who imprint a much deeper mark and accelerate the development of the ecosystem. This effect is hard to measure using traditional outcome proxies, but it is a significant element in the narrative, sensemaking and community-supporting behaviours of local entrepreneurs.

The rise of entrepreneurial ecosystems in the Balkans represents a change of mindset. The process is irrevocable, even though most affluent business people in the region still invest in old-fashioned extractive businesses rather than in technology-driven entrepreneurial ventures. However, due to the educational effect of the young entrepreneurial ecosystems on society, there is a younger population group that approaches life and work differently and that has a global rather than a parochial view about what they do and its potential impact. The idea of sharing and giving back to a community without a fee or a rent has been rather revolutionary in the post-Cold War period, as materialistic and overly individualistic cultural elements flourished in the years after the fall of Communism.

Entrepreneurial ecosystems in the Balkans offer many opportunities for reflection and learning. In several Balkan countries, where xenophobia, racial discrimination and related intolerance ignited recent conflicts, there still exists a real need for diversity to enable and enrich the further development of the entrepreneurial ecosystems. National pride, to the extent it exists among the entrepreneurial communities, is not a destructive force, and entrepreneurial communities appreciate the opportunity to work with and learn about others.

Entrepreneurial finance has accelerated the development of the local ecosystems, and it acted as a cohesive force for the region. The JEREMIE fund of the European Commission and the European Investment Fund channelled through the Bulgaria-based accelerator venture funds made possible the creation of an incipient pan-Balkan entrepreneurial community. The fact that Bulgaria became the centre of regional entrepreneurship is explained by the fact that it was able to access a significant pocket of money, which enabled entrepreneurial experimentation on a much larger scale, access to training and numerous events that brought global superstars and mentors closer to the regional entrepreneurs.

Local entrepreneurs seek inspiration from the Zuckerbergs and the Musks of this world in places such as Silicon Valley, London or Berlin. There is generalized ignorance about the entrepreneurs who have succeeded in the Balkan region under similar circumstances and in the face of similar challenges that are found throughout the region. This is partly due to the limited linkages between the countries, which leads to poor sharing of success stories and lessons learned. We see this as a huge missed opportunity in terms of knowledge exchange and mutual learning within the region.

Entrepreneurship means hope and optimism for the region. The development of entrepreneurial ecosystems is seen as a mechanism for solving the structural institutional issues in the region, as entrepreneurs take up active civil positions and participate in political life as well as in the NGO sector. Moreover, technological entrepreneurship, which is predominant in the ecosystems, is not generally subject to corruption and provides regional entrepreneurs with an opportunity to enter global markets, thereby overcoming the burden of weaker regional institutions.

In essence, participants describe entrepreneurial ecosystems in the Balkans as arrangements that give rise to successful entrepreneurs and firms, producing recognizable success stories that offer relevant examples and role models for at least some part of the population. Entrepreneurial ecosystems consist of communities with an established culture of giving, of active citizenship, abounding with aspirations to change the world. These entrepreneurial ecosystems provide a sense of purpose and meaning for individuals and communities with problem-solving, experimentation-oriented attitudes. Entrepreneurial ecosystems with a significant IT component might be hard to track and account for because their activities are located in multiple centres (locally, but also in big client centres such as London, Boston, etc.). This tendency to have multiple locations, together with the intentional obscuring of country of origin because of anticipated negative country-related associations, and the fact the early stages of ecosystem development involve much networking and community building make the traditional measures of the ecosystem environment inappropriate and even potentially damaging. They can curb the optimism and the enthusiasm of nascent and early-stage entrepreneurs and reduce the morale of the community. For this reason, the celebration of early positive developments is important, and universities, the media and NGOs play a very important role here.

As authors, we have written this book as a much-needed and well-deserved tribute to all those who have been contributing to entrepreneurship as a means of bringing prosperity and optimism to the region and who have started a shift in the mindset of significant population groups. We hope that our readers will see this book as a manifesto for the entrepreneurial community—written whilst in constant communication with its participants—and one that fairly documents its diversity, the early signs of maturity in the local ecosystems and the agglutinating power of the entrepreneurial mindset. As such, the book is an organic product of the community and is an integral part of a deep process of cultural change that has and will have profound effects on the region's economic, social and political realities. This 'softer' side is complemented by data-rich analyses that draw on both secondary and primary data collection efforts in which the authors of the book were personally and directly involved. The argument is based on hard data, detailed case studies and more than 70 interviews and is framed as a rigorous academic research endeavour.

References

Acs, Z., Stam, E., Audretsch, D., & O'Connor, A. (2017). The lineages of the entrepreneurial ecosystem approach. *Small Business Economics, 49*(1), 1–10. https://doi.org/10.1007/s11187-017-9864-8

Auerswald, P. (2015). *Enabling entrepreneurial ecosystems*. Kauffman Foundation Research Series on City, Metro, and Regional Entrepreneurship, Ewing Marion Kauffman Foundation.

Borissenko, Y., & Boschma, R. (2016). A critical review of entrepreneurial ecosystems: Towards a future research agenda. *Paper in Evolutionary Economic Geography, 16*(30), 1–20.

Brown, R., & Mason, C. (2017). Looking inside the spiky bits: A critical review and conceptualization of entrepreneurial ecosystems. *Small Business Economics, 49*(1), 11–30.

Eleven quarterly report. (2017). Private correspondence with the managing founders.

Feldman, M., & Braunerhjelm, P. (2007). The genesis of industrial clusters. In *Cluster genesis: Technology-based industrial development* (vol. 1, pp. 1–13). Oxford University Press.

Isenberg, D. (2011). The entrepreneurship ecosystem strategy as a new paradigm for economic policy: Principles for cultivating entrepreneurship. The Babson Entrepreneurship Ecosystem Project, Babson Global. Retrieved from http://entrepreneurial-revolution.com/2011/05/11/the-entrepreneurship-ecosystem-strategy-as-a-new-paradigm-for-economic-policy-principles-for-cultivating-entrepreneurship/

Livingston, J. (2007). *Founders at work: Stories of startups' early days*. Berkeley, CA: Apress.

Martin, R. (2015). Rebalancing the special economy: The challenge of regional theory. *Territory, Politics, Governance, 3*(3), 235–272.

Mason, C., & Brown, R. (2014). *Entrepreneurial ecosystems and growth oriented entrepreneurship* (pp. 1–38). OECD.

Merriam-Webster dictionary online. Retrieved from https://www.merriam-webster.com/dictionary/balkanized

Neumeyer, X., & Corbett, A. (2017). Entrepreneurial ecosystems: Weak metaphor or genuine concept? In *The great debates in entrepreneurship* (pp. 35–45). Published online: August 10, 2017. Permanent link to this document: https://doi.org/10.1108/S1048-473620170000027005

Schmutzler, J., Andonova, V., & Diaz-Serrano, L. (2018). How context shapes entrepreneurial self-efficacy as a driver of entrepreneurial intentions: A multilevel approach. *Entrepreneurship Theory and Practice*. https://doi.org/10.1177/1042258717753142

Stam, E. (2015). Entrepreneurial ecosystems and regional policy: A sympathetic critique. *European Planning Studies, 23*(9), 1759–1769.

The Entrepreneurship Database Program at Emory University, summary report from 2016. Retrieved from https://www.galidata.org/assets/report/pdf/2016_Year_End_Summary.pdf

Zahra, S., & Nambisan, S. (2012). Entrepreneurship and strategic thinking in business ecosystems. *Business Horizons, 55*, 219–229.

2

Entrepreneurial Ecosystems in the Balkan Peninsula

Introduction

Silicon Valley and Berlin are not the only places where successful new ventures thrive. Great business ideas pop up and grow in unlikely places—in communities that manage to leverage business savviness, talent and creativity. In Europe, the entrepreneurial ecosystems east of Berlin are nascent, but they have already gained breadth and depth and show potential.

Between 2010 and 2017, 5 billion euros were generated as returns to investors in Central and Eastern Europe (CEE) thanks to 15 Venture Capital-backed exits, seven of which were from the Balkan region. This has resulted in cities like Sofia becoming the most important hubs for growth for the tech sector in CEE, creating incentives for firms to set up funds there (Ezekiev, 2017). These recent developments are not just luck; they are a logical consequence of the vibrancy of the regional entrepreneurial ecosystems which, whilst not free from problems, offer a huge learning opportunity about how to enable entrepreneurial ecosystems in unlikely places.

V. Andonova et al., *Entrepreneurial Ecosystems in Unexpected Places*,
https://doi.org/10.1007/978-3-319-98219-9_2

Why Would a Chinese Peroxide Company Pay US$1 Billion for a Slovenian Talking Feline?

Samo and Iza Login, Slovenian entrepreneurs who started the company Outfit7, created one of the most popular apps in recent times (Satariano, 2017). Talking Tom is an app franchise that is No. 1 in more than 100 countries on the App Store and has more than 350 million monthly active users. The game is highly profitable, utilizing both advertising and in-game purchase offers. Outfit7 earns more than US$100 million a year.

In 2016, the creators decided to sell the company, and they found a buyer in Zhejiang Jinke Peroxide Co., a Chinese industrial chemical maker with revenues of US$133 million in 2016. Outfit7 was sold for US$1 billion. Although this acquisition might look strange on paper, given that the synergy between a chemical maker and a game company is not obvious, this is one of several cases where industrial Chinese companies have tried to expand into the entertainment business to the benefit of Balkan entrepreneurial ventures. Of course, some might argue that is the unusual nature of the Chinese economy that makes such operations possible. According to CODE Advisors LLC, an investment bank that specializes in media and technology deals, 70% of all acquisitions of game companies since 2015 have been by Chinese buyers (Satariano, 2017). In addition, the Chinese stock market has unusual ways of operating, leading to companies being able to acquire targets with high growth potential at high prices in a quest for improved profitability. Because regulators and investors in China focus almost exclusively on a company's bottom line, industrial companies with low margins have started to bet on game studios and entertainment companies that have better margins to increase their stock price. However, because of China's strict regulations on outward investment, these mergers and acquisitions can be very complicated and obscure.

Western companies do not face these limitations in their transborder acquisitions, and they do not want to miss out on attractive acquisition targets from the Balkans either. For example, on 22 October 2014, the US-based Progress Software acquired Telerik, a Bulgarian software company founded in 2002 by Vassil Terziev, Svetozar Georgiev, Boyko Iaramov and Hristo Kosev, for US$262 million. Telerik offers software tools for web, mobile and desktop application development and tools and subscription services for cross-platform applications. They have received multiple awards for their products and business practices. One of the most important awards was the 2013 Best of TechEd Award for Software Development. Telerik in 2018 had offices in Bulgaria (HQ), the USA, Australia, Germany, India, Denmark and the UK. The company's rapid growth and expansion was accomplished through the acquisition and through mergers with other overseas companies, including the buying of Vanatec GmbH from Germany (2008), a merger with ArtOfTest in the USA (2010) and the acquisition of TeamPulse from Canada (2010), Fiddler from the USA and EQATEC Analytics from Denmark (2013). These acquisitions had the purpose of offering support to additional areas of the development process of software and its lifecycle.

Key Entrepreneurship Indicators

In order to assess the state of development of the Balkan entrepreneurial ecosystem, this book uses multiple data sources, providing a comprehensive picture of the state of development and maturity of different elements of the Balkan entrepreneurial environment. We do this from a comparative perspective, either across countries and regions or longitudinally, comparing different time periods in order to take full advantage of available but still limited data about the entrepreneurial ecosystems in the region.

We use the key entrepreneurship indicators provided by the Adult Population Survey (APS) for the Global Entrepreneurship Monitor (GEM) as aggregate performance proxies of entrepreneurial ecosystems. Academic teams from each economy that participates in the GEM data collection effort administer and oversee this survey, which is conducted using a random representative sample of at least 2000 adults between the ages of 18 and 64 years. The surveys are conducted at the same time every year (between May and July) using a standardized questionnaire provided by the GEM Global Data Team. The questionnaire is translated into local languages, and back-translated for a validity check.

The key entrepreneurship indicators built into the information collected by the APS comprise all the phases of the lifecycle of a business: nascent entrepreneurship rate, new business ownership rate, Total Early-stage Entrepreneurial Activity (TEA), established business ownership rate and business discontinuance. The latest available results are reported in Table 2.1. Invariably, the Balkan economies show low rates of business ownership (but also lower business discontinuance). Their TEA rate is almost twice as low as those of the Baltic countries and is almost two percentage points lower than that of Central European countries. These differences are significant, and they have to be explained and reconciled with the upbeat statements about entrepreneurship coming from the region. With this objective in mind, we will dig deeper into the details and tap into multiple information sources. Certainly, the traditional and well-established ways of measuring entrepreneurship fail to capture significant ongoing developments that are not reflected in the outcome proxies available until 2018. So, what do we really know about Balkan entrepreneurial ecosystems?

Table 2.1 Total entrepreneurship activity reported in the Global Entrepreneurship Monitor Report of 2017

	Nascent entrepreneurship rate	New business ownership rate	TEA	Established business ownership rate	Business discontinuance
The Balkans	4.08	2.7	7.64	8.2	5.92
Baltic states	10.7	4.85	15.2	8.65	8
Central Europe	5.27	4.17	9.37	6.23	8.47

Balkans represented by Bulgaria, Greece, Macedonia, Slovenia and Croatia
Baltic states are represented by Estonia and Latvia
Central Europe is represented by Hungary, Poland and Slovakia
Data retrieved from 2017 Global Entrepreneurship Monitor Report

As a point of departure, we use the National Experts Survey (NES) of the Global Entrepreneurship Monitor (GEM).

The Entrepreneurial Framework Conditions of the GEM

Every year, each economy participating in the GEM cycle interviews at least 36 key experts and informants, who have been previously identified as knowledgeable and desirable sources of opinion (see GEM annual reports, various years). In this regard, the National Experts Survey (NES) is similar to other surveys that capture expert judgements to evaluate specific national-level conditions. However, the NES focuses only on the environmental features that are expected to have an important impact on the entrepreneurial activities, captured in nine different but very specific entrepreneurial framework conditions. The nine entrepreneurial framework conditions as identified by GEM include financing for entrepreneurs, government policies, governmental programmes, entrepreneurial education and training, research and development transfer, commercial and professional infrastructure, internal market openness, physical and services infrastructure and social and cultural norms.

When all the country data is collected, the files are harmonized centrally by the GEM data team, which includes an internal quality audit and the calculation of site variables that summarize each block of questions designed to measure a certain aspect of the entrepreneurial climate. Each year, different countries take part in the GEM survey, and the results are compared across countries. When it comes to the Balkans, the yearly country comparison is a real challenge because very few countries have a sustained commitment to measuring essential elements of their national entrepreneurship climate. Despite these data limitations, the available measures provide insights.

An important advantage of the NES questionnaire is that it is standardized for all countries, which makes the data readily comparable. It is carefully designed and refined to capture informed judgements from national key informants in each country, who are selected on the basis of their reputation and experience. Experts are asked to express their views about the most significant conditions either fostering or constraining entrepreneurial activity and the development of the entrepreneurial ecosystem in their country. The National Experts Survey, therefore, provides insights into the ways in which essential elements of the entrepreneurial ecosystems of Balkan countries shape the regional entrepreneurial climate and determine the future productive base of their regional economies.

The experts are interviewed using a semi-structured questionnaire. The closed questionnaire consisted of several statements relating to aspects of the nine entrepreneurial framework conditions. The responses are measured using a Likert scale of 1 (highly insufficient) to 5 (highly sufficient). The data obtained from the respondents is analysed in order to determine the mean score for each category of questions. In Table 2.2

Table 2.2 Entrepreneurial framework condition scores as reported in the National Experts Survey of GEM, weighted average (1 = highly insufficient, 5 = highly sufficient)

Measurement	The Balkans	Baltic states	Central Europe
Financial environment for entrepreneurship	2.28	2.83	2.435
Specific governmental support and entrepreneurship priority	2.30	NA	2.25
Government policies: taxes and bureaucracy	2.33	2.77	2
Government entrepreneurship programmes	2.47	3.03	2.385
Entrepreneurship education: primary and secondary level	2.23	2.74	1.635
Entrepreneurship education: vocational, professional and tertiary level	2.92	3.02	2.575
R&D transfer	2.49	2.60	2.045
Access to commercial and professional infrastructure	2.95	3.37	2.88
Internal market dynamics	2.46	2.84	3.435
Internal market burdens	2.50	2.82	2.68
Access to physical infrastructure/services	3.45	4.13	3.75
Cultural and social norms	2.48	3.23	2.36

Balkans represented by Romania, Macedonia and Slovenia
The Baltic countries included here are Estonia, Latvia and Lithuania
The Central European countries included here are Poland and Slovakia
Data retrieved from the National Experts Survey in the Global Entrepreneurship Monitor Report, 2013

we present historical results comparatively, using the Baltic countries and several Central European countries for which data is available as a benchmark group.

According to the national experts, the Balkans have a number of significant comparative weaknesses (those with a score below 2.5), especially in comparison to the Baltic countries. The most critical ones have to do with internal market dynamics and burdens that, in many cases, have to do with favouring incumbents. Market dynamics refers to the extent to which markets change dramatically from year to year, and market openness is the extent to which new firms are free to enter existing domestic markets. The other key weakness of the Balkan entrepreneurial climate has to do with dominant social and cultural norms. These existing social and cultural norms have traditionally discouraged rather than encouraged individual actions that might lead to new ways of conducting business or economic activities, which might, in turn, lead to greater dispersion of personal wealth and income.

The most important strengths of the Balkan economies in relation to the entrepreneurial framework conditions as captured by the traditional measures are access to physical infrastructure and services and access to commercial and professional infrastructure. The former includes ease of access to available physical resources—communication, utilities, transportation, land or space—at a price that does not discriminate against new, small or growing firms. The latter includes the presence of commercial, accounting and other legal services and institutions that allow or promote the emergence of small, new and growing business entities.

Government Programmes and the Public Sector

Government programmes and policies and the functioning of the public sector are important factors in entrepreneurial ecosystems and are the most important drivers in turning entrepreneurial intentions into actual entrepreneurial behaviour. Even though in modern market economies it is not the government that starts new businesses, government policies and initiatives can shape the conditions that are conducive to entrepreneurial endeavours. Government policies related to their support of entrepreneurial ecosystems are rated by Balkan national experts in Table 2.3.

The governments of the Balkan states are given credit for imposing taxes at levels that do not place a huge burden on new and growing firms. In addition, local governments in the region appear to show interest in making entrepreneurship a priority. Still, despite both of these relatively positive accounts, the region appears to lag considerably behind the government policy ratings of the Baltic states. At the same time, regulations and bureaucracy are among the most burdensome aspects of government-related policies to impact on entrepreneurship. In Table 2.4, we report the most recently available data (2017) on three of the most critical activities governments can contribute to with the objective of supporting the development of national entrepreneurial ecosystems: support and

Table 2.3 Average expert ratings for government policies as reported in the National Experts Survey, weighted average (1 = completely false, 5 = completely true)

Statement	The Balkans	Baltic states	Central Europe
Government policies consistently favour new firms.	2.02	2.43	1.86
Support for new and growing business is a high priority for policy at national government level.	1.88	2.93	1.92
Support for new and growing business is a high priority for policy at local government level.	2.38	2.57	1.91
New firms can get most of the required permits and licences in about a week.	1.73	2.67	1.97
The amount of taxes is NOT a burden for new and growing firms.	2.21	2.33	2.2
Taxes and other government regulations are applied to new and growing businesses in a predictable and consistent way.	1.65	3.17	1.94
Coping with government bureaucracy, regulations and licencing requirements is not unduly difficult for new and growing firms.	1.65	2.77	1.67

Balkans represented by Romania, Macedonia and Slovenia
The Baltic countries included here are Estonia, Latvia and Lithuania
The Central European countries included here are Poland and Slovakia
Data retrieved from the National Experts Survey in the Global Entrepreneurship Monitor Report, 2013

Table 2.4 Perceived support by government and quality of policies reported in the Global Entrepreneurship Monitor Report of 2017

	Government policies and support		
	Support and relevance	Taxes and bureaucracy	Entrepreneurship programmes
The Balkans	3.14	3.34	3.56
Baltic states	4.45	4.75	4.7
Central Europe	3.4	3.03	3.57

Balkans represented by Bulgaria, Greece, Macedonia, Slovenia and Croatia
Baltic states represented by Estonia and Latvia
Central Europe represented by Hungary, Poland and Slovakia
Data retrieved from 2017 Global Entrepreneurship Monitor Report

relevance of government programmes; taxes and bureaucracy; and state-sponsored entrepreneurship programmes. On all these accounts, the Balkan states are ranked consistently lower than the Baltic countries and on a par with Central European countries.

More specifically, as Table 2.5 shows, in 2017, it takes on average 6 procedures, 16 days and almost 4% of the per capita income of residents to start a business in the Balkans. As a comparison, in the Baltic States, it takes 4 procedures, 5 days and 1% of per capita income.

There is certainly an increased number of government programmes for startups and entrepreneurial ventures, as well as initiatives related to the creation of science parks and business incubators, which, in the case of Bulgaria, was led by the highest-ranked public official—the president

Table 2.5 World Bank Doing Business 2017—Starting a Business (average results for sample groups)

	The Balkans		Baltic states		Central Europe	
Indicator	Mean	Std. dev	Mean	Std. dev	Mean	Std. dev
Procedures—number	6.11	2.62	3.67	0.58	6.5	1.29
Time—number of days	16	19.21	4.83	1.15	16.38	13.94
Cost—% of income per capita	3.68	4.26	1.2	0.6	4.88	5.17

Source: The World Bank. Ease of Doing Business 2017 report
Balkan countries include Bulgaria, Romania, Greece, Macedonia, Serbia, Albania, Croatia, Slovenia and Bosnia Herzegovina
Baltic states include Lithuania, Latvia and Estonia
Central European countries include Hungary, Poland, Slovakia and the Czech Republic

of the country. There are, however, multiple ways in which the workings of these facilities as well as the state apparatus might be improved and transformed into powerful support mechanisms for the entrepreneurial community.

A significant number of key stakeholders in the Balkan entrepreneurial ecosystem expressed concerns about the ability of state officials to carry out even the most thoughtfully designed programmes for entrepreneurial support. For example, in Bosnia and Herzegovina, the multiple levels of government put into place with the objective, in part, to guarantee the transition from war to peace in a multi-ethnic setting is seen as a major problem for the development of a more vibrant entrepreneurial community, not so much because of the associated bureaucratic processes but because of the high opportunity cost that abundant and well-paid government jobs impose on young potential entrepreneurs. On the other hand, in Bulgaria, the fact that many public servants were incapable of understanding the business models of some of the young foreign market-oriented technology ventures was presented as a fortunate circumstance! Entrepreneurs felt they were protected from extortion by the ignorance of government officials rather than by the institutions for law enforcement in the country. In other countries, such as Serbia, where only private sector money fuels the entrepreneurial ecosystem, corruption is seen as a marginal problem for the entrepreneurial community because it presents itself mostly where public funds are distributed. The absence of government financing implies absence of corruption, and entrepreneurial ventures that have emerged and flourished under even the harshest market conditions show remarkable resilience.

Serious reservations about the capacity of government officials to effectively and competently carry out their tasks are pervasive across the region. Among the best-rated aspects of government entrepreneurship programmes in the Balkans is the support offered by science parks and incubators, where there is significant involvement by the private sector and successful entrepreneurs, who participate as mentors, role models and investors. In general, better coordination with the private sector and the incipient entrepreneurial community can be used to improve the talent pool and the efficiency of existing government programmes designed to stimulate entrepreneurial endeavours. Following this line of argument, government efforts in countries like Bulgaria to establish programmes for new and growing businesses are recognized by the experts as better than most of the other government initiatives.

Infrastructure, Human Capital and Innovation Capabilities

Entrepreneurial ecosystems thrive in places with well-developed infrastructure. National experts assess physical infrastructure as being among the most competitive elements that contribute to successful entrepreneurial ecosystem development in the Balkans, as evident from the data presented in Table 2.2. The infrastructure indicator in the Global Innovation Index for the region shows a wide margin for improvement in this regard, however.

The Global Innovation Index provides a framework and assessment metrics for the performance of 127 economies around the world, systematizing evidence and offering a broad vision built on 81 indicators, including infrastructure. The infrastructure pillar comprises of measures of information and communications technologies (ICTs), general infrastructure and ecological sustainability. Table 2.6 contains comparisons between the average scores of Balkan, Baltic and Central European countries. Even though differences in the infrastructure score between the

Table 2.6 Global Innovation Index 2017—Infrastructure Pillar (average score and ranking for sample groups)

	The Balkans		Baltic states		Central Europe	
Infrastructure	Mean	Std. dev	Mean	Std. dev	Mean	Std. dev
Score	49.08	6.47	58.07	5.45	54.55	2.22
Avg. rank among 127 economies	56	21.6	28	15.72	38.25	6.6

Source: The Global Innovation Index 2017 Report. Infrastructure Pillar

Balkan countries include Bulgaria, Romania, Greece, Macedonia, Serbia, Albania, Croatia, Slovenia and Bosnia Herzegovina

Baltic states include Lithuania, Latvia and Estonia

Central European countries include Hungary, Poland, Slovakia and the Czech Republic three regions remain relatively small, the average global ranking of the Balkan region is significantly lower than the best performing Baltic region: the Balkans are placed at 56, while the Baltic countries as a group are placed at 28. In essence, physical infrastructure in the region still has a long way to go despite pockets of excellence: Romania and Bulgaria feature among the top 30 countries in the world with the fastest internet connection (The Sofia Globe, 2017). Still, according to the EU's Digital Economy and Society Index (DESI), in 2018 Romania comes last and Bulgaria comes 26th from among the 28 EU member states. This is because the human capital dimension of the index is the most critical: in both countries, the overall level of digital skills is amongst the lowest in the EU, and it varies significantly between different socio-economic groups. The best performing Balkan country amongst the EU members is Slovenia, which shows slightly above-average performance when compared to other EU members.

Extant research has shown that there is a correlation between the levels of perceived capabilities of would-be entrepreneurs, entrepreneurial intentions and the level of total early-stage entrepreneurial activity (Eurostat, 2012; Tsai, Chang, & Peng, 2016). Therefore, education is inextricably linked to entrepreneurial intentions and the vibrancy of the entrepreneurial ecosystem, as it affects entrepreneurs' confidence in having the necessary skills and knowledge to start a business. In essence, education that is focused on entrepreneurship-related skills can have a powerful influence over entrepreneurial intentions and behaviour.

Entrepreneurial experience and/or education help develop new skills that can be applied to many spheres in life. Non-cognitive and soft skills, such as opportunity recognition, innovation, critical thinking, resilience, decision-making, teamwork and leadership, benefit all economic spheres, whether or not individuals intend to become or continue as entrepreneurs, as they lead to more sophisticated decision-making at all levels and in every type of private or public organization. More specifically, there is evidence that practical entrepreneurship training may better prepare school leavers for the transition from school to the labour market, enabling them to identify business opportunities and improving their chances of success in business and self-employment ventures (Cheung & Chan, 2011). Entrepreneurship education can enhance an individual's level of self-confidence, as well as increase their interest in entrepreneurship as a viable career choice.

Universities and Entrepreneurship in the Balkans

Universities are key players in the startup ecosystem because they hold and attract young talent; shape and influence students' mindset; create and serve as a repository of knowledge and expertise in learning and education, all of which nurture entrepreneurial ecosystems. At their best, universities can commit to the support of entrepreneurial ecosystems if they inspire proactiveness and promote a culture of innovation. They can help students strengthen the required skills and knowledge they require as entrepreneurs and act as a stage for experimentation and early failure, a platform for research in support of the ecosystem and a place for sharing knowledge and best practices. In sum, as an indispensable part of the most developed entrepreneurial ecosystems in the world, universities engage in lasting relationships with ecosystem partners to build synergies for continuous improvement (Andonova & Nikolova, 2015).

In the Balkans, however, universities have contributed little to local entrepreneurial ecosystems. Even though, occasionally, STEM-related departments will receive some credit for their ability to provide large-scale basic training, the majority of actors in the Balkan entrepreneurial ecosystem perceive university processes and their knowledge base as irrevocably antiquated and out of sync with the requirements of present-day economies. Qualifications are scarcer in the case of state-owned universities, where the introduction of new and innovative programmes is perceived as a threat to the long-lasting and cosy status quo of faculty members who are government employees; as a result, new programmes are either kept small-scale or aborted altogether soon after the pilot launch. The incentives for universities and their staff to be entrepreneurial and innovative are just not present.

The vacuum left by the short-sightedness of the educational sector in the Balkans is huge, and there are multiple private initiatives that are addressing the unfulfilled need for adequate entrepreneurial skills. NGOs, such as Junior Achievement; development agencies, such as Swisscontact; and private academies, such as the Telerik Academy are among the most active educational partners on the Balkan entrepreneurial scene. The case of Telerik Academy is particularly significant. Telerik Academy was established in 2009 to offer free education to people of all ages on behalf of the Bulgarian IT industry. Candidates were screened for basic abilities and aptitudes and were expected to show a long-term commitment to pursuing a career in the IT industry in Bulgaria. The training programme lasted less than a year and consisted of several modules, culminating with a significant client project supervised by an experienced Telerik programmer. The best and brightest (top 5–10% of all graduates) were given job offers at Telerik, a highly valued opportunity in the local labour market, while the rest went to other employers in the industry. In this way, Telerik Academy was effectively able to cream off the best talent, but it also provided much-needed free training and educational services for the local economy. Unfortunately, this significant contribution to the local entrepreneurial ecosystem was not perceived as pivotal for the business once Telerik was acquired by Progress in 2014.

The patience and resilience with which organizations such as Telerik Academy have sustained the regional entrepreneurial ecosystem is admirable, but the remaining gap is too big to be satisfactorily filled by the private sector. An ambitious and thorough educational reform that transforms the university sector from its roots upwards is urgently demanded by the remaining actors in the Balkan entrepreneurial ecosystem—investors, entrepreneurs, managers of established companies and NGOs. If the Balkan entrepreneurial ecosystem is going to produce its first unicorn in the near future, it needs all its cogs to be in place. One reason for optimism is that local talent is not in short supply.

Talent Is Abundant

Vlad Tenev, a native Bulgarian who moved to the USA as a child, started his professional career in the financial industry before building several startup companies, mostly unsuccessful fintech ventures (Mararakis & Shontell, 2017). With a partner, he founded Robinhood, an app that makes stock trading simple and efficient, removing fees and many other barriers that have prevented the mass adoption of stock trading; it was worth US$1.3 billion in 2017.

The company blew up overnight when their website was promoted on a famous tech news website, and 1 million people signed up for the app, almost a year and a half before its release. The entrepreneurs had to raise around US$3 million before they could launch the app due to regulatory practices, and many people saw it as too risky. Tenev claims they got rejected by over 75 venture capitalists. Others, such as Marc Andreessen, the American entrepreneur and investor, took a bet on the product. Thanks to access to a privileged network, Tenev was able to get investors such as Snoop Dogg. In 2017, over 2 million people used the app. It operates using a freemium model, that is, the basic app is essentially free, but people pay for extra features. Robinhood won the 2015 Apple Design Award. Most of its users are under 30, and it has helped people learn how to invest in stocks and make the experience easy and efficient. Usually, investing money on the stock market would require a broker, paying of fees and waiting for extended periods of time for transactions to go through. With Robinhood, people are able to invest small amounts of money instantly for free, getting access to financial instruments they would not have been able to use in the past. By tracking their transactions in the app, investors have been able to make money and to save.

The regional programming talent of the Balkans is recognized worldwide. According to data from Stack Overflow, the dominant Q&A platform for coders with about 3.5 million users, the best coders are from the Balkans (Salkever, 2015). Stack Overflow ranks the skill level of coders around the world by taking into account the 'up or down' votes to answers to previously posted questions about coding and systems, as well as the ranking of fellow users. As such, the result is entirely crowd-driven. Looking at the average country rankings of 14,898 Stack Overflow users who have a reputation score of 5000 or more, Bulgarian coders come at the top, scoring the highest average reputation in the world. Bulgaria also has 40 top performing users of Stack Overflow, and their impressively high scores make Bulgaria the country with the highest average reputation in the world, while Croatia and Greece, the other Balkan countries in the top 20, appear in 6th and 19th places, respectively.

This top performing talent in coding and other spheres comes at a rather modest price when we consider the average wages, salaries and benefits paid by entrepreneurial ventures in the Balkans. Table 2.7 contains a comparison between the average yearly personnel expenses of Balkan, Baltic and Central European early-stage ventures, pointing to huge gap in the average cost of labour, although top spenders in both regions report comparable maximum expenditures (US$266,505 vs. US$250,000). On average, entrepreneurial ventures from the Balkans operate at a 50% discount when compared to labour costs in the Baltic states and on a par with companies from Central Europe. However, entrepreneurial ventures from the Balkans have a much higher share of founders with technical degrees as the highest educational achievement, and also have a much more academically focused founders' profile than Central European ventures.

The educational profiles of new venture founders in the Balkans and in the Baltic states are as shown in Table 2.8. The average number of years of education completed by the most highly educated founder is 17 for Balkan entrepreneurial ventures and 18 years for Baltic ventures, with a maximum of 21 years in both cases. The percentage of founders who have PhD degrees as the highest level of education completed by the most highly educated founder is twice as high for Baltic entrepreneurial ventures when compared with Balkan entrepreneurial ventures, whose founders more frequently have bachelor or technical degrees as the highest level of education as shown in Table 2.9.

Table 2.7 Average yearly wages, salaries and benefits paid to workers in US dollars (2013–17, US$)

	Average	Standard deviation	Minimum	Maximum	Sample size
The Balkans	$12,518	$43,759	$0	$266,505	48
Baltic states	$23,784	$68,521	$0	$250,000	12
Central European states	$10,305	$23,763	$0	$92,000	19

Balkan states represented are Albania, Bulgaria, Romania, Serbia, Greece, Croatia, Macedonia and Slovenia
Baltic states represented are Estonia, Latvia and Lithuania
Central European States represented are Hungary, Poland and Slovakia
Data provided by the Entrepreneurship Database Program at Emory University, supported by the Global Accelerator Learning Initiative
Data collected between 2013 and 2017

Table 2.8 Highest level of education completed (in years) by the founder with the most education (2013–17)

	Average	Standard deviation	Minimum	Maximum	Sample size
The Balkans	17.00	2.41	12	21	24
Baltic states	17.75	2.36	16	21	7
Central European states	13.80	3.05	6	16	15

Balkan states represented are Bulgaria, Romania, Serbia, Greece and Slovenia
Baltic states represented are Estonia, Latvia and Lithuania
Central European states represented are Hungary, Poland and Slovakia
Data provided by the Entrepreneurship Database Program at Emory University, supported by the Global Accelerator Learning Initiative
Includes author's calculations. Data collected between 2013 and 2017

Table 2.9 Highest level of education completed by the founder with the most education (2013–17)

Level of education	The Balkans states (%)	Baltic states (%)	Central Europe states (%)
PhD	10	20	0.0
Masters	41	40	26.7
Bachelor	15	30	26.7
Technical degree	23	0	33.3
High school	10.30	10	6.7
Middle school	0	0	6.7

Balkan states represented are Albania, Bulgaria, Romania, Serbia, Greece, Croatia, Macedonia and Slovenia
Baltic states represented are Estonia, Latvia and Lithuania
Central European states represented are Hungary, Poland and Slovakia
Data provided by the Entrepreneurship Database Program at Emory University, supported by the Global Accelerator Learning Initiative. Includes author's calculations
Data collected between 2013 and 2017

Entrepreneurial Finance

Across CEE, venture capital investment amounted to €0.5 billion in 2015, representing 3% of the total amount invested in the venture capital field across Europe in that year (Ezekiev, 2017). Both private and institutional money shows increasing interest in supporting regional ventures in this part of the world and in the Balkans, more specifically.

In this context, Sofia emerged strongly as an accelerator capital in Europe (#3 in 2017 according to the president of BVCA, Evgeny Angelov). In the years after 2012, there was EU funding for early-stage startups in the form of seed and venture capital. The European Union invested about US$1.5 billion in Bulgaria under a programme called Joint European Resources for Micro to Medium Enterprises (JEREMIE). A total of US$150 million of this investment along with money from the Bulgarian government was put under the supervision of the European Investment Fund which selected four funds as the managers of a share of this amount conditional on their ability to raise additional private funds. The total amount was around €350 million, revolving around innovation funding to support startups and SMEs. These investments accomplished a 2.57 multiplier effect which then mobilized a total of €875 million in financing. This has enhanced the competitiveness of 9476 SMEs and 180 startups, helping to support more than 25,550 jobs in the region (Angelov, 2017). According to the European Investment Fund, growth has been exponential: in 2016, 210 startups raised US$74 million up from 20 companies and US$4 million (O'Brien, 2018).

Two accelerator venture funds were instrumental in these developments—Eleven and LAUNCHub—as well as others such as Neveq, a venture equity fund that has traditionally focused on revenue-generating companies rather than startups. Eleven was a US$15 million startup accelerator and seed fund, while LAUNCHub started as a US$12 million pre-seed fund. Ultimately, most of the entrepreneurial finance came from the EU, a situation that invites possible policy discussions on the role of local governments as an arms-length limited partner investor. The experience in the Balkan countries shows that the government needs to focus more on creating a broad context for the functioning of the entrepreneurial ecosystem but cannot be an active manager of the funds, as the right incentives lie with the partners and managers, not bureaucrats. There are still challenges, and there are market gaps in both equity and debt instruments in the region and a marked scarcity of investment funds beyond the seed level. In many countries from the region—in Romania, for example—funding for early-stage ventures is still very bureaucratic and as a result totally unprepared for the speed of response that these startups need in order to survive and thrive.

The region also faces a lack of late-stage funding opportunities which creates incentives for promising local entrepreneurial ventures to sell prematurely, as they do not expect to be able to access readily available funds to fuel their growth. This, however, is a dilemma all entrepreneurs in young ecosystems might like to have as it is a strong indicator that the base of the ecosystem is solid and set for expansion.

Eleven, LAUNCHub and the Accelerator Experience

The two accelerator venture funds that started operations in 2012 were instrumental for Balkan entrepreneurial community building. This process was catalysed by the financial resources that both accelerators were able to channel, attracting substantial private capital. Private money brought discipline to companies and investments, and it also brought expertise and know-how. The newly built community that formed around these two loci of the Balkan entrepreneurial landscape provided support with contacts, scrutiny of strategy and keeping the ventures and their portfolios on track. Thus, there were many additional benefits apart from the money that supported the creation of a community. Establishing a culture of belonging and giving back was probably one of the most important among these.

The hugely important educational role that these two accelerator venture funds needed to perform in order to inform and develop Balkan entrepreneurs can be understood from the data available from the Entrepreneurship Database Program at Emory University. Tables 2.10 and 2.11 show the relative importance of the benefits that the Balkan and the Baltic entrepreneurs, respectively, sought in their participation in 178 accelerator and support programmes around the world. The complete database as of February 2018 surveys a total of 13,495 ventures. According to this data, Balkan and Baltic entrepreneurs sought mostly network development. However, Baltic entrepreneurs recognized business skill development and mentorship to be essential elements of the accelerator experience, in contrast to the Balkan entrepreneurs who gave a higher priority to securing direct funding and connections to potential investors. This evidence may indicate two things: the relative scarcity of venture

Table 2.10 Relative importance of accelerator benefits as perceived by entrepreneurs—Balkan states (2013–17)

Benefit importance—The Balkans

Rank	Network development (%)	Business skills development (%)	Mentorship (%)	Connection to potential investors (%)	Secure direct funding (%)	Access to a group of entrepreneurs (%)	Awareness and credibility (%)
1	34	19	11	11	19	4	2
2	17	11	11	36	13	6	6
3	11	17	30	11	23	9	0
4	19	15	15	19	11	6	15
5	11	17	17	17	15	17	6
6	6	9	15	2	13	36	19
7	2	13	2	4	6	21	51

Balkan states represented are Albania, Bulgaria, Romania, Serbia, Greece, Croatia, Macedonia and Slovenia
Data collected between 2013 and 2017. Data provided by the Entrepreneurship Database Program at Emory University, supported by the Global Accelerator Learning Initiative

Table 2.11 Relative importance of accelerator benefits as perceived by entrepreneurs—Baltic states (2013–17)

Benefit importance—Baltic states							
Rank	Network development (%)	Business skills development (%)	Mentorship (%)	Connection to potential investors (%)	Secure direct funding (%)	Access to a group of entrepreneurs (%)	Awareness and credibility (%)
1	36	27	18	9	0	0	9
2	36	9	18	9	18	0	9
3	9	27	27	18	9	9	0
4	9	9	9	27	9	27	9
5	9	27	0	18	36	9	0
6	0	0	9	18	18	27	27
7	0	0	18	0	9	27	45

Baltic states represented are Estonia, Latvia and Lithuania. Data collected between 2013 and 2017

Data provided by the Entrepreneurship Database Program at Emory University, supported by the Global Accelerator Learning Initiative capital in the Balkan region and a marked overconfidence among Balkan entrepreneurs in their business skills and abilities. Both arguments seem plausible. As one of the leading entrepreneurs from the region put it: 'The entrepreneurial ecosystem especially for tech ventures suffers from far too much optimism and far too little knowledge.' He might be right, but what is certain is that entrepreneurship needs optimism and the region has not been traditionally known for the optimistic views of its inhabitants. Something certainly has changed, and reasons for optimism are not lacking. In February 2018, BGmenu, an online food delivery startup in the LAUNCHub portfolio, announced it had been acquired for almost US$13 million by Takeaway.com, a Netherlands-based company. This was a much-needed success for a young entrepreneurial ecosystem.

Looking for Regional VC Investors

Successful VC players are active in the region, and many of them are homegrown. Take the case of Romanian Gecad Ventures, which successfully exited Vector Watch and RAV Antivirus, the latter sold to Microsoft in 2003. Still, the overall level of VC investments in Romania, the largest Balkan country in terms of population, is just 0.001% of GDP, compared to the EU average of 0.027%, according to Invest Europe (MacDowall, 2017). In Croatia, there is a single fund, South Central Ventures, that also covers the West Balkans. There is also a group of angel investors, but they are not very active. In Romania, for example, startup ventures are also starved of capital, and the most promising of these relocate their core management operations to the UK or the USA in search of venture financing.

On average, however, regional financing needs have remained comparatively low. Most of the startups are in the ICT, digital, engineering and mid-tech sectors because the regional ecosystem promotes this kind of startups, given the tradition of successful innovations and the presence of corporations, education and community organizations with such a focus. The entry cost and capital requirements in these industries are low to moderate, and the region already has infrastructure and a history of success in these domains, allowing startups to leverage complementary resources, thus limiting their capital requirements and the need for entrepreneurial finance.

According to the application data obtained from Eleven—the accelerator venture fund—between November 2012 and December 2015, 81% of startups in the region reported they had invested their own funds in their startup ventures, with an average amount invested equal to US$20,210. The average amount sought by ventures to grow their operations, according to LAUNCHub data, between 2014 and 2015, was US$645,730, a modest amount.

Market Functioning and Private Sector Sophistication

The entrepreneurial landscape in the Balkans is mostly defined by its success in turning Eastern and Central Europe into hubs for digital and tech startups, some of which have been globally recognized despite the fact they did not become household names. The region has some well-recognized advantages, such as low tax rates. In Bulgaria, for example, there is a 10% flat tax on profits and 5% on dividends. This, plus the growing number of well-run co-working places, accelerators and VCs, is setting the scene for the Balkans to be seen as one of the best regional entrepreneurial ecosystems. There are some unquestionable challenges too: the fragmentation and variety of the small markets within the region is frequently perceived as one of the biggest obstacles that startups in this area face for future growth (Guerrini, 2017).

Startup Internationalization in the Balkans

The presence of leading global companies with R&D operations in the Balkans has unquestionably improved the knowledge and relevant experience available in the region's labour markets. Nevertheless, regional buyers are not very sophisticated, and often there is no local demand for the products that the advanced regional engineering talent creates. This fact, more often than not, pushes entrepreneurial ventures into early-stage internationalization, adding another layer of complexity to the operations of these ventures. Two significant challenges arise from the fact that local markets are small and unsophisticated: first, Balkan entrepreneurs experience a shortage of business skills when it comes to internationalization because the large majority of them lack proper business training; second, in the not too distant past, the mostly negative image of Balkan countries produced a negative spillover effect, even for the most innovative entrepreneurial ventures. Some of the region's most successful B2B ventures were explicitly asked to keep the names of their world-renowned clients confidential for fear that the Balkan origin of their suppliers might invite suspicion and mistrust about their operations. This has been changing gradually, largely because some of the most successful regional entrepreneurial ventures made substantial efforts to educate their clients and show pride in their country as part of their business strategy. Branding the region is very important for the future of Balkan entrepreneurial ecosystems, and there has been an exponential growth in awareness and efforts in this regard. Getting recognized as an attractive place to live and work by citizens and foreign talent is essential for the further maturing of the young regional ecosystems. Regional governments, especially those of Balkan EU member countries, are taking cautious steps to encourage the immigration of high-skilled workers and attract talent from non-EU member neighbours. More frequently, there are voices in favour of more aggressive government programmes of high-skilled immigration from outside the region. If there is a wide consensus between actors within the ecosystem, it is that for a vibrant entrepreneurial ecosystem the most important thing is the people, the passion and the commitment to do something different.

Bitdefender from Romania is an excellent case in point. In 2017, Bitdefender was ranked by PricewaterhouseCoopers as the 17th largest emerging market software company, and the company claims to serve more than 500 million users. This global cybersecurity firm began as an entrepreneurial effort by Florin Talpeş and his wife Mariuca just after the end of the Communist regime. Early entrepreneurs and other businesses in the region were threatened by cyber criminals and computer viruses. This situation led Talpeş to design antivirus programs in order to protect his startup, which focused on outsourcing software to the French market due to the political and economic proximity of these two countries. France was one of the first countries to invest in and trade with Romania in the technology and car components sectors, establishing several assembly lines for computers and cars in the country. Bitdefender's successful products were then requested by other clients starting new businesses, which led to a global expansion in 2001, quickly reaching other big markets such as the USA, Germany, Great Britain, Canada and Australia. The company's innovations in fighting cyberattacks were a breakthrough in the industry, and it launched a novel technological solution that was rapidly embedded in security software globally. Led by Bitdefender, Romanian cybersecurity solutions were considered one of the world's best, becoming a byword in cybersecurity innovation. With the development of this sector and the accession of the country to the European Union in 2007, which led to an improvement in the business climate and opportunities in Romania, this Romanian brand was starting to be recognized internationally, creating a sense of trust and reliability that encouraged venture capital firms to invest in the country. The new entrepreneurial ecosystem that has developed between 2012 and 2018 enabled Romanian startups to grow faster than they did in the 20 years following the fall of Communism. Bitdefender represents a model of a homegrown business that employs hundreds of Romanians and does not look to relocate its core management operations overseas. Bitdefender certainly had a solid base upon which to build their business.

Multinationals and the Labour Market

Within the Comecon system of the Soviet bloc countries, Romania and Bulgaria experienced economic specialization in information technology. This IT tradition, coupled with a strong STEM-focused education, which, however, has declined since the fall of the Berlin Wall in 1989, gave both Bulgaria and Romania a strong start in the digital economy when they became members of the EU in 2007. Bulgaria rapidly became a competitive outsourcing destination of leading incumbents in the digital economy, such as Cisco Systems, Hewlett-Packard, VMWare, Microsoft, Oracle, SAP and IBM. The presence of the IT leaders and the availability of advanced IT professionals prepared the ground for IT-based entrepreneurial development. According to the president of the Bulgarian Venture Capital Association, the export of IT-related products and services has grown more than four-fold since 2008. These powerful companies, however, also impose a heavy burden on the Balkan entrepreneurial ecosystem because they compete against domestic entrepreneurial ventures for the same local talent.

Between 1990 and 2007, when Bulgaria joined the EU, 1 million Bulgarians left the country, an average of 60,000 per year. According to estimates made in 2017, 30,000 leave each year, most of whom are students seeking graduate degrees in IT, engineering and medicine (Hope, 2018). Immigration from the Balkans to Western Europe has been an ongoing process since the early twentieth century. As recently as 2015, more than 130,000 migrants from Kosovo, Albania and Serbia sought asylum in the EU (Zeneli, 2017). Their main motivations have been to look for better paid jobs in richer countries and escape from institutional despair (*The Guardian* Editorial, 2015). The Balkan countries have suffered from poor governance and corruption, which pushes talent to seek opportunities in the wealthier countries of Western Europe and the USA. The small size of the Balkan nations makes the impact of the 'brain drain' even greater, as this reduces private sector activity, productivity and the overall competitiveness of the region.

The Global Competitiveness Report from the World Economic Forum 2016/2017 ranked Serbia 137th out of 138, Bosnia 134th and Croatia 132nd, with Albania and Macedonia slightly ahead for 'Capacity to retain talent'. This reflects dramatically the number of young students that reported their intentions to leave the country after graduating, including a notable 85% of Macedonian students. In Serbia, in 2016, more than 58,000 people left, doubling the previous yearly average. The high level of youth unemployment further motivates people to leave the region looking for better opportunities in other countries. In Kosovo, for example, 60% of unemployment relates to young people, while in Croatia it is 45% (DW, 2016). This situation is sometimes seen as a vicious cycle, where young talent from the Balkan region leaves, searching for better opportunities, and this further harms the economic situation, living standards and business productivity.

The ongoing structural problems affecting the functioning of the market in the Balkans is sometimes seen as an ongoing transitioning from the Communist period that ended in the early 1990s. Most of the economies in the Balkans are small and open. There has been some progress, with the transition to working market institutions from the start of the 2000s that introduced structural reforms and market stabilization which has led the Balkan economies to improve their attractiveness to foreign investors. The EBRD transition index for the region—a measure of the progress of countries in transition towards a free market economy—rose from an average of 2.3 in 2000 to 3.2 in 2008 (Izvorski, 2015). One aspect that is expected to radically improve the working of markets and help all Balkan economies develop is joining the EU. As of 2018, Albania, Macedonia, Serbia and Montenegro were candidates for admission to the EU, but there is no firm commitment as to their accession date.

Macedonia, for example, has suffered from difficulties in developing free markets, mostly due to problems with cartelization and corruption which, to a varying degree, affect the entire region. There have been large and recurrent corruption scandals in the Macedonian government since 2015 (*The Economist*, 2016a). However, there have also been remarkable advances in terms of foreign investment in the country. Since 2017, the government has directed its efforts towards lowering taxes and attracting investors with incentives. These efforts contributed to positioning the country among the top ten in the World Bank's Ease of Doing Business Ranking for 2016 (Byrne, 2017). Although there is still a substantial margin for improvement, regional governments have a clear agenda for promoting market-based relationships, and their regulatory initiatives have been internationally acclaimed.

Culture

One defining element of Balkan culture is that pessimism is all-pervasive. Even the most progressive and supportive members of the Balkan entrepreneurial ecosystem agree that the transition in mentality from its Communist past has been the hardest, in part because the post-Communist governments in the region have not made a complete transition from the old ways of thinking and doing. A generation change is expected to bring about a much-needed shift in mentality, especially outside the already global community of IT and high-skilled professionals and managers which has developed in this part of the world.

Given the recent Communist past of the countries in the region, it is not surprising that there is a very strong orientation towards employment rather than self-employment. Participants in the local entrepreneurial ecosystem describe the tendency to shy away from responsibility and the over-analysing of failure as yet another cultural constraint for the faster development of regional entrepreneurial ecosystems. The business culture is incipient, and this can also be seen in the consumption patterns of local tycoons, who spend lavishly on yachts and other luxury goods but do not see themselves as investors in the local entrepreneurial ecosystem.

Occasionally, the sense of entitlement found among young and overconfident entrepreneurs in Balkan entrepreneurial ecosystems repels investors. Albanian startup founders, for example, are perceived as inexperienced and underprepared but also arrogant (Bohanes, 2017). They have been accused of a lack of planning and trust in their attitudes towards investors, causing reduced investment in their startups. Hence, it is of critical importance that entrepreneurs strengthen their reputation for trustworthiness with all participants in the entrepreneurial ecosystem but particularly with investors. This could be rapidly achieved with improved business skills; a thorough knowledge of the business fundamentals and indicators; evidence of efforts to do the utmost to make the business a success; evidence of a disciplined approach; a willingness to learn; the appropriate behaviour for business settings; driven and resourceful attitudes and an appreciation of the trust required to be recommended to other investors.

Members of the diaspora community and returnees are also important participants in the regional ecosystem, bringing a significant cultural shift to the region. Even though there are no readily available statistics about the number of returnees in the Balkan entrepreneurial ecosystem or about the diaspora members who are actively supporting Balkan-based ventures, the names of the more prominent actors are widely known within the community. According to the National Statistical Institute and the Economic Institute with the Bulgaria Academy of Sciences, there are more Bulgarians working abroad than in the country, and 55% of those who emigrate every year are aged 20–29 (O'Brien, 2018). Those who really commit to and work hard in this ecosystem have a completely global outlook, and they participate in global networks, which imprints a pattern of behaviour that makes Balkan entrepreneurs indistinguishable from their global partners.

Media

The essential role of the mass media in creating role models and promoting entrepreneurship is unquestionable. One of the weaknesses within the region is the lack of sufficient celebration of the stories of success or entrepreneurship more generally (even if ending in failure). The general media pays some attention to entrepreneurship, but the power of social media is unstoppable, and entrepreneurial communities in the Balkans have leveraged mainly social media to become stronger.

Many laws and policies regarding a free media have been established, but the proper application of these has been difficult and is not very effective, because of the concentration of the media market. There are frequent accusations that media freedoms are restricted due to political pressure from government, subsidies which result in political bias and intimidation and violence against journalists. Most of these accusations can be traced back to one and the same cause: a problematic media ownership that is often subject to personal relations and interdependencies and the influence of political power. This furthers the problem of transparency, making it hard for new media outlets to enter the market without entering into political power struggles (Brogi, Dobreva, & Parcu, 2014). A lack of proper education for journalists and low journalistic ethical standards in the region makes this problem worse. This, coupled with a high level of perceived corruption in the region, makes the media unreliable, as many established journalists are often exposed to a high level of internal and external pressures that results in limited media freedom (Bieber & Kmezic, 2015). Moreover, the protection offered to journalists is very low. According to the Human Rights Watch in 2016, in 86 interviews conducted with journalists in the region, there was ample evidence of intimidation and violence, especially against those who report on war crimes, religious extremism and corruption (*The Economist*, 2016b).

Nevertheless, from an informational point of view, the media now reports more news about entrepreneurship and technology, and there is more information available on how to build an entrepreneurial venture regardless of where you live, and the Balkans are no exception. Capital has started to flow to this part of the world, whose citizens are noted for their sometimes rebellious nature, an essential ingredient for a functioning entrepreneurial ecosystem. However, as yet, local entrepreneurs are still not seen as a positive force for societies in the region, in particular by leading intellectuals. Entrepreneurs and intellectual elites in this part of the world still look at each other with suspicion and mistrust.

Academic and Research Institutions and R&D Transfer

Public spending on R&D in the Balkans is relatively low compared to other European countries. Economies like Bulgaria, Croatia, Greece and Serbia spent around 0.9% of their GDP on R&D in 2015. In contrast, on average EU countries spent 2.05% of their GDP on R&D in the same year, while Baltic and Central European countries such as the Czech Republic, Hungary and Estonia spent around 1.5–2% on R&D. Romania is far behind, with public spending on R&D for 2015 hovering at around 0.5% of GDP. In the Balkans, the positive outlier is Slovenia, which spent more than the average rate for the European Union, at 2.2% of GDP (The World Bank, 2016).

Innovation capabilities—which are important for an economy's ability to become competitive, particularly in higher-productivity sectors—are heavily dependent on research and development. Innovation capabilities require a business environment that in its turn facilitates entrepreneurship and provides access to the necessary finance for the creation and growth of innovative firms. Such an environment needs to be supported by effective university and research institutions, with strong links to industry and an ability to integrate with local industrial clusters. It is clear that participants in the Balkan entrepreneurial ecosystem believe that universities do not play a central role in facilitating knowledge transfer and stimulating innovation, and that the virtual circle between university-based innovation, entrepreneurship and competitiveness in the region is broken. Key stakeholders in the regional entrepreneurial ecosystem believe that the root cause of the problem is the broken linkage between the academic and research entities and the market.

While there is a consensus that better integration between R&D centres and industry is welcome, there is no consensus about whether lavish public spending on R&D that gives rise to entrepreneurialism is desirable. Take the case of one of the champions of the startup nations—Israel. The pharmaceutical Teva is the only multinational success in the Israeli economy headquartered in that country which has raised the question whether the high level of investment made by the government into research and development has sufficiently benefited the country, given the recent turn to low-margin generic drugs that has profoundly damaged its profitability, as well as the profitability of the Israeli pension funds that contributed to its early success. In fact, hundreds of Israeli technology startups have been snapped up by global firms and moved abroad, denying the opportunity for the establishment of a much more lucrative sector in the economy, leading to the realization that there are two economies: one lucrative and high-tech that benefits a few and one that is much less thriving but where most workers find employment (*The Economist*, 2017). Still, it is the high-tech, exponential-growth type of entrepreneurship that is at the core of Balkan entrepreneurial ecosystems.

Technological Entrepreneurship

The growth of technological entrepreneurship in Western Europe and in the Balkans, in particular, is a proof of the resilience and promise of the latter. A few decades ago, people used to link technology with Silicon Valley in the USA. Nowadays, however, important technological startups have been formed all around the world, and Europe in particular has seen tremendous growth of tech startups.

In 2018, Europe's technological landscape is large and well established, and its regional entrepreneurial ecosystems are growing rapidly due to the capability for innovation found on the continent. There is also a positive mindset among young Europeans towards entrepreneurship. Furthermore, Europe has a large pool of talent in the tech industry. According to Stack Overflow Insights, there were 4.7 million professional developers in Europe in 2016, compared to 4.1 million in the USA. This talent is also being supported by world-class academic institutions, with five of the top ten Computer Science schools in the world being European, including ETH in Zurich and Oxford in England. Furthermore, one in five MBA graduates from the top-ranked business schools in Europe is entering the tech industry.

The tech industry in Europe is clustered in hubs all over the continent (The State of European Tech, 2017). The largest hubs are located mostly in Western and Northern Europe in cities such as London, Paris, Berlin, Stockholm and Helsinki. But many others are growing rapidly, including cities in Eastern Europe like Warsaw, Prague and Budapest. All these tech hubs and entrepreneurship opportunities mean tech is creating more jobs than other industries. In 2016, the European tech industry employment growth was averaged at 2.1% compared to a 1.0% increase in total employment growth forecast by the European Commission for the same time period. In the tech industry, countries like Ireland showed a 5.5% growth in tech worker population, with several others showing a 2–3% growth for 2016 (The State of European Tech, 2017).

Ultimately, the UK, Germany and France are emerging as the main destinations for tech talent to emigrate and start ventures. This is mainly due to entrepreneurs looking for the best talent available and for a high-quality ecosystem with the best conditions in which to start their venture. This is especially true in cities such as London and Berlin, where the tech ecosystem is more mature and advanced, compared to other hubs. As of 2018, however, there is more access to capital, not only within the largest hubs, but all over Europe. There are more investors looking all over the continent for good opportunities, regardless of the traditional hubs. This has helped entrepreneurs and investors to follow the talent instead of the money, which has resulted in more successful ventures across the region, giving a strong edge to the Balkans as a talent-rich region. In 2016, half of all the tech-related meetups occurred outside of the top 20 hubs. From 2015 to 2016, several Eastern European cities, including Prague, Belgrade, Vilnius and Sofia, have at least doubled their number of tech meetups. This is where we start to see countries like Bulgaria, Serbia, Romania and others from the Balkan region emerging as potential tech hubs within Europe, building and consolidating the different elements of their entrepreneurial ecosystems.

According to Slush & Atomico's State of the European Tech Report in 2016 (The State of European Tech, 2017), a number of cities in the Balkans were already identified as emerging tech communities. Cities like Sofia (Bulgaria), Belgrade (Serbia) and Bucharest (Romania) have seen some of the largest levels of growth from 2015 to 2016. Sofia has seen the largest rate of growth, from 13 meetups in 2015 to 33 in 2016. Belgrade had a similar jump, from 13 to 28. The largest by far is Bucharest, which had almost three times as many meetups at 90, yet still doubled its total from 2015. This trend suggests technological entrepreneurship in the Balkan region is on an accelerated trajectory and is expected to grow even more in the coming years.

The overall dynamics of the entrepreneurial ecosystems in the Balkans, as measured by the Global Entrepreneurship Monitor, tend to show a very positive trend between 2016 and 2018, too. In terms of perception among the population, results show that there are more perceived opportunities in the countries and lower fear of failure rates. Perceived opportunities have gone from 18.2% in 2016 to 22.96% in 2018, slowly closing the gap with Europe, where 43.4% of the population identify positive entrepreneurial opportunities. At the same time, the fear of failure rate has gone from 36.75% in 2016 to 32.4% in 2018, which is lower than the European average of 37%. At the same time, entrepreneurial motivation in the Balkans has become more opportunity than necessity-driven, pointing to the potential of higher value-added entrepreneurship. The region has started to develop its potential, but the action still is happening elsewhere.

In 2017, most of the deep tech startups and ideas still came from the main tech hubs of the continent, mainly the UK, France and Germany, with significant developments in areas such as artificial intelligence, virtual reality and augmented reality (The State of the European Tech

Report, 2017). Tech giants such as Apple, Amazon and Google are focusing on deep tech within Europe, investing in these same countries. At the same time, innovation levels among early-stage entrepreneurs in the region which can be directly related to technology have not changed much between 2016 and 2018. The region still lacks a level of innovation comparable to other parts of Europe, but considering its turbulent historical and socio-political past, the developments are more than impressive.

The Entrepreneurial Ecosystem in Romania

Like all post-Communist Balkan countries, Romania's entrepreneurial ecosystem has a short history, which only began to develop over the last 30 years. This late start is mostly attributed to the political and economic regime that had prevailed in the country until 1989, which focused on state ownership and control of every political, economic and social activity. Thus, the national environment did not stimulate any kind of private initiative or business skills development, and professional education emphasized strong technical abilities. During this period, entrepreneurs were not encouraged, but computer technology became one of the most developed industries with high levels of investment in research and development as part of the political programme of the ruling regime. Among the countries of the Communist bloc, Romania and Bulgaria were the most focused on information technology and became leaders in computer software and hardware.

The fall of the Communist regime brought about a new social landscape, and a more benign environment for new ventures in Romania. Many startups in different industrial sectors succeeded, and some of them eventually became national champions in banking, online retail, medical services, electronics retail, internet provision and security software development. Nonetheless, as some of these companies sought internationalization, they faced significant challenges. Romania's international image in the technology sector in particular was not good: the country's Communist past and the damage done by hackers operating in the country contributed to an image of unreliability and risk. In order to advance with their entrepreneurial ventures, Romanian entrepreneurs had to convince investors and clients they were trustworthy by establishing a brand that had a positive international reputation. This has been a recurrent criticism of Balkan entrepreneurial communities, as they lack the skills to build internationally renowned local brands.

In the post-Communist years, the Romanian economy has suffered from a lack of specialization, although it had developed several internationally competitive pockets of economic activity in retailing, business services, healthcare, automotive components, communication equipment and IT. A significant obstacle to the development of a more consolidated entrepreneurial ecosystem has been a lack of entrepreneurial financing. Apart from this limitation, which has been masterfully surmounted in Bulgaria, for example, the outlook for the Romanian entrepreneurial ecosystem is promising. Clearly, it will benefit greatly from increased business skills education and entrepreneurship experience, and it would be energized by the occasion of successful exits. Nevertheless, it is hard not to notice a certain disappointment among entrepreneurial ecosystem participants following significant international acquisitions, as multinational corporations do not show the same level of involvement and contribution to the development of the local ecosystem as the national entrepreneurial ventures and their founders had.

Looking to the Future

The future of the nascent entrepreneurial ecosystems in the Balkans is clearly set. These economies have walked away from outsourcing and entered into a phase of building genuine entrepreneurial ecosystems, whose growth is a positive sign for all classes of stakeholders within the region. They understand the advantages and disadvantages of their institutional and historical context, geographical location, talent pool and cost drivers. They have pride in and are engaged with their national and regional context while having a strong global mindset. For the most part, they are determined to make their dreams come true, engaging in civil society and political movements in order to drive forward a faster pace of positive change. Is this wishful thinking? We do not think so. Even though there might be an element of fad and fashion, with 'entrepreneuring as a lifestyle' acquiring a very attractive status, in particular for younger generations, Balkan entrepreneurial ventures are built for the most part on very solid business and economic logic, low-cost advantages and very difficult market conditions due to the small size of the markets and the price sensitivity of local clients. The business logic is strong, while local culture is biased towards pessimism rather than optimism, balancing the natural overoptimism of most entrepreneurial founders, and the technical education of the founders is strong. Most importantly, the connecting tissue of these ecosystems—the entrepreneurial communities—has started to function as such. The notion of giving back and the idea of being a part of a network to which entrepreneurs contribute but from which they also benefit have already crystallized. The focal points around which these networks form do not have as much to do with country boundaries as they do with sources of essential resources for entrepreneurism, such as finance and talent. The Balkan region is very diverse, and some generalizations will certainly apply more to some rather than to all entrepreneurial ecosystems. However, the drive to think and present the region as such rather than as a collection of individual countries has been getting stronger support from the stakeholders within the Balkan entrepreneurial ecosystems. These developments are not captured by traditional measures of entrepreneurship, but they are a very real part of the ongoing processes and mindset shifts in this part of the world. The entrepreneurial ecosystems of the Balkans provide a new type of dynamics for the region, far removed from the logic of violence, nationalism and separation but one which involves community building, meritocracy and a consciousness of the interdependencies between business and societal wellbeing from a wider regional and even global perspective. Few people associate the Balkans with these sentiments, but they had better get used to it!

References

Andonova, V., & Nikolova, M. (2015). *Universities & entrepreneurship. Why universities are important members of the start-up ecosystem*. Presentation, Sofia, Bulgaria.

Angelov, E. (2017). *Funding entrepreneurs*. Bulgarian Venture Capital Association, Presentation, Vienna. Private communication with the author on May 31, 2017.

Bieber, F., & Kmezic, M. (2015). *Background paper: Media freedom in the Western Balkans*. Balkans in Europe Advisory Group. Retrieved January 20, 2018, from http://www.biepag.eu/wp-content/uploads/2016/10/BIEPAG-Media-Freedom-in-the-Western-Balkans.pdf

Bohanes, M. (2017). Albanian entrepreneurs: For heaven's sake, leave your entitlement at the door! [Blog post]. Retrieved July 25, 2017, from http://www.oficina.al/albanian-entrepreneurs-for-heavens-sake-leave-your-entitlement-at-the-door/

Brogi, E., Dobreva, A., & Parcu, P. L. (2014). *Freedom of media in the Western Balkans*. European Parliaments Subcommittee of Human Rights, EU, Belgium. Retrieved November 9, 2017, from http://www.europarl.europa.eu/RegData/etudes/STUD/2014/534982/EXPO_STU(2014)534982_EN.pdf

Byrne, A. (2017). What put Macedonia in World Bank's top 10 ranking for investors. *The Financial Times*. Retrieved March 7, 2018, from https://www.ft.com/content/4f786fa0-3b18-11e7-ac89-b01cc67cfeec

Cheung, C., & Chan, R. (2011). The introduction of entrepreneurship education to school leavers in a vocational institute. *International Journal of Scientific Research in Education, 4*(1), 8–16.

DW. (2016). Youth are deserting Balkan countries. Retrieved January 26, 2018, from http://www.dw.com/en/youth-are-deserting-balkan-countries/a-36891266

Entrepreneurship Database Program at Emory University. (n.d.) Retrieved from https://docs.wixstatic.com/ugd/4d837d_797f386156d042128e88da78cb5ad30c.pdf

Eurostat. (2012). *Entrepreneurship determinants: Culture and capabilities*. European Union. Retrieved June 2, 2018, from http://ec.europa.eu/eurostat/documents/3217494/5748437/KS-31-12-758-EN.PDF

Ezekiev, P. (2017). The $5 billion CEE exits lead table. *Neo Ventures* [Blog Post]. Retrieved February 15, 2018, from https://neoventures.net/2017/08/06/5-bn-lead-table-cee-rising-in-technology/

GEM Annual Reports Various Years. (n.d.). Retrieved June 13, 2018, from https://www.gemconsortium.org/report

Guerrini, F. (2017). Is Sofia the real digital capital of the new markets? *Forbes*. Retrieved February 24, 2018, from https://www.forbes.com/sites/federicoguerrini/2016/04/14/is-sofia-bulgaria-the-real-digital-capital-of-the-new-markets/print/

Hope, K. (2018). Bulgaria battles to stop its brain drain. *FT*. Retrieved March 3, 2018, from https://www.ft.com/content/51f1bd86-d6cc-11e7-ae3e-563c04c5339a

Izvorski, I. (2015). The three transitions of the Western Balkans. *The World Bank Blog*. Retrieved April 19, 2017, from http://blogs.worldbank.org/developmenttalk/three-transitions-western-balkans

MacDowall, A. (2017). How Romania became a popular tech destination. *Financial Times*. Retrieved January 28, 2018, from https://www.ft.com/content/a0652dba-632f-11e7-8814-0ac7eb84e5f1

Mararakis, A., & Shontell, A. (2017). The founders of Robinhood, a no-fee stock trading app, were initially rejected by 75 venture capitalists—Now their startup is worth $1.3 billion. *Business Insider*. Retrieved July 6, 2017, from http://www.businessinsider.com/robinhood-app-vlad-tenev-founder-free-stock-trading-valuation-2017-7

O'Brien, C. (2018). Bulgaria rising: Can a growing startup movement reinvent the country's economy? Retrieved May 19, 2018, from https://venturebeat.com/2018/03/23/bulgaria-rising-can-a-growing-startup-movement-reinvent-the-countrys-economy/

Satariano, A. (2017) Why did a Chinese peroxide company pay $1 billion for a Talking Cat? *Bloomberg Businessweek*. Retrieved May 17, 2017, from https://www.bloomberg.com/news/features/2017-05-17/why-did-a-chinese-peroxide-company-pay-1-billion-for-a-talking-cat

The Economist. (2016a). Protesters want to topple Macedonia's Government over a corruption investigation. Retrieved January 18, 2018, from https://www.economist.com/europe/2016/04/21/protesters-want-to-topple-macedonias-government-over-a-corruption-investigation?zid=307&ah=5e80419d1bc9821ebe173f4f0f060a07

The Economist. (2016b). Media freedom is dwindling in the Balkans. Retrieved January 18, 2018, from https://www.economist.com/europe/2016/11/08/media-freedom-is-dwindling-in-the-balkans?zid=307&ah=5e80419d1bc9821ebe173f4f0f060a07

The Economist. (2017, August 10). Start up and leave nation: An Israeli pharma champion sickens. Retrieved August 14, 2017, from https://www.economist.com/news/business/21726083-country-frets-about-what-tevas-decline-means-israeli-pharma-champion-sickens?frsc=dg%7Ce

The Guardian Editorial. (2015). The Guardian view on the EU and migration: Let's stop the Balkans being part of the problem. Retrieved June 22, 2017, from https://www.theguardian.com/commentisfree/2015/aug/27/the-guardian-view-on-the-eu-and-migration-lets-stop-the-balkans-being-part-of-the-problem

The Sofia Globe. (2017). Internet connection speeds Bulgaria among top 30 in the world. Retrieved February 7, 2018, from https://sofiaglobe.com/2017/08/09/internet-connection-speeds-bulgaria-among-top-30-in-the-world-2/

The State of European Tech. (2017). Retrieved May 25, 2018, from https://2017.stateofeuropeantech.com/

The World Bank. (2016). Research and development expenditure (% of GDP). Retrieved from https://data.worldbank.org/indicator/GB.XPD.RSDV.GD.ZS?locations=EU&name_desc=false

Salkever, A. (2015). Data: Best programming talent in the world is not in California. Retrieved May 19, 2018, from https://venturebeat.com/2015/04/05/data-best-programming-talent-in-the-world-is-not-in-california/

Tsai, K. H., Chang, H. C., & Peng, C. Y. (2016). Refining the linkage between perceived capability and entrepreneurial intention: Roles of perceived opportunity, fear of failure, and gender. *International Entrepreneurship and Management Journal, 12*, 1127. https://doi.org/10.1007/s11365-016-0383-x

Zeneli, V. (2017). Migration, brain drain and the Western Balkans. Retrieved February 11, 2018, from https://www.theglobalist.com/european-union-migration-brain-drain-and-the-western-balkans/

3

The Creation and Development of the Ecosystem: The Role of ICT

Introduction

While the Balkan countries do not hold the lead when it comes to innovation, to some extent entrepreneurship is part of the traditional mindset and culture. In Southern and Eastern Europe, the self-employment rate and share of micro and small enterprises contributing to local economies tend to be higher than in Northern and Western European countries (Hatfield, 2015). This suggests that starting one's own business comes more naturally to these geographic regions than many other parts of Europe because of the structure and type of economic activities. In Greece, for example, SMEs are a true backbone of the economy and of these four out of five are sole proprietor companies—a much higher share compared to the average one in every four for Europe (Mylonas and Tzakou-Lambropoulou, 2016).

The content in this chapter is largely grounded in the insights and opinions of the many members of the regional startup community interviewed in the context of this book project. Some of this material, especially in the context of historical developments, is used to supplement the facts and insights available from public sources and literature. Many of

V. Andonova et al., *Entrepreneurial Ecosystems in Unexpected Places*,
https://doi.org/10.1007/978-3-319-98219-9_3

their observations have been synthesized to extract key insights about factors and facts that are fundamental to the regional entrepreneurial dynamic.

Current Entrepreneurship Dynamics and Key Players in the Region

The recent rise of entrepreneurship across the Balkan region results from a combination of regionally specific factors that align with the global rise of entrepreneurship as a natural process of modernization of contemporary economies and advancing societies. In Europe especially, there has been an overall shift towards entrepreneurship facilitated by government policies that enable the creation and commercialization of new knowledge and new business models (Audretsch et al., 2000). In that sense, it is natural that the entrepreneurial communities and startups of the region share some common features with their peers across more advanced and entrepreneurial economies but that they also carry some specific features.

As in many other places, the technological startup community consists mostly of well-educated younger people who have some experience of working with larger successful multinational companies or who have gathered experience abroad, mostly while working or studying in Western Europe or North America. The majority of startups that serve as inspiration and role model to others in the region are in the technology domains, either providing products or services geared towards the global IT industry or providing technology-optimized solutions for other industries. However, there are still active discussions throughout the startup communities in the region about what entrepreneurship is and what the true characteristics of an actual entrepreneur are. Should an entrepreneur in this region be different from an entrepreneur in Western Europe, North America or anywhere else? Figure 3.1 summarizes some of the most common background characteristic of the Balkan entrepreneur who is an active participant in the regional entrepreneurial community.

There is no question that startup communities of different sizes exist across all countries in the region, but they are of different size and maturity. In Bulgaria, Croatia, Slovenia and Romania, they are larger, more

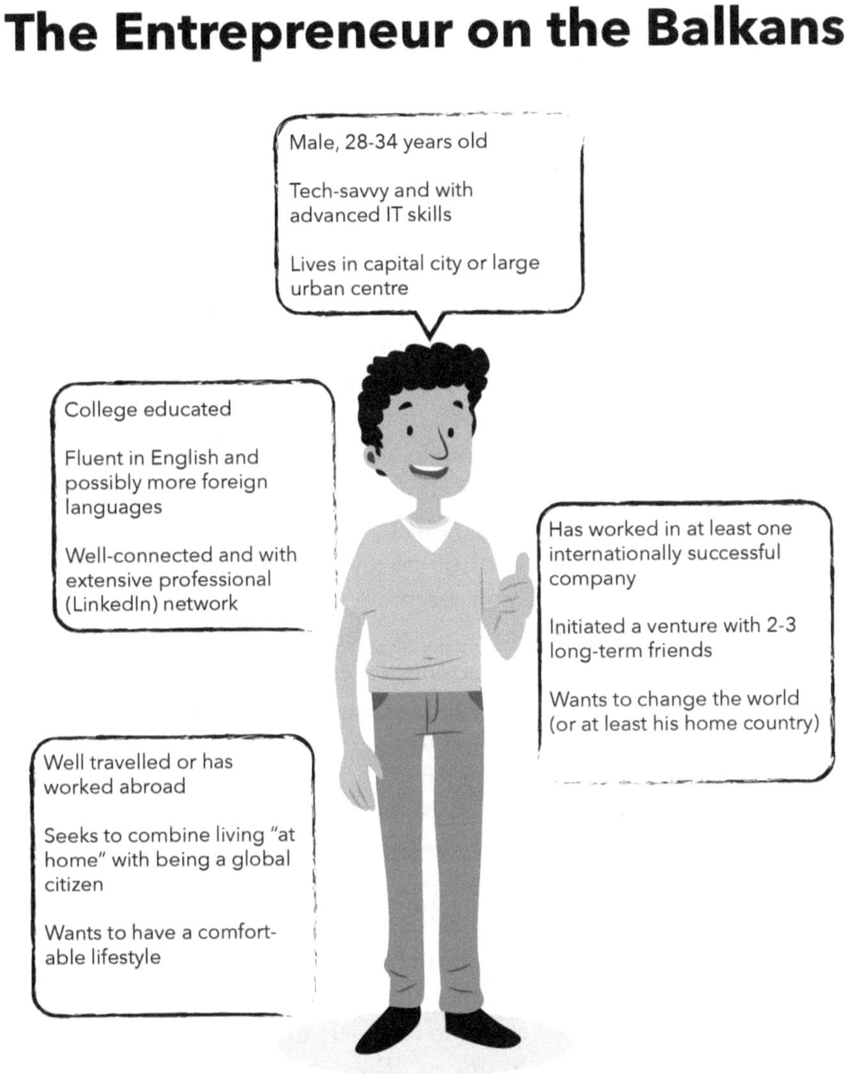

Fig. 3.1 The entrepreneur in the Balkan region

influential and more mature compared to those in some of the neighbouring countries. Contemporary entrepreneurial processes in these countries, all of which are members of the European Union, started earlier than in the rest of the region and have benefited from the overall

EU trend of nurturing startup-friendly socio-economic environments. Greece, which is also a member of the EU, remains a bit of an oddball in the region when it comes to entrepreneurship. While it is the oldest EU member in the region and is one of the countries with the strongest traditions in individual entrepreneurship and has the highest share of small businesses (Mylonas and Tzakou-Lambropoulou, 2016), it does not produce great examples when it comes to encouraging contemporary technology-based entrepreneurship. Despite its ability to access EU funds and benefit from the experience of fellow EU economies, Greece is home to only a small and rather loose startup community with few impactful success stories. One of the anecdotal explanations for the fact that it has been surpassed by some of the younger EU members in the region is that while there is a strong local entrepreneurial culture, mostly in the leading local economic sectors such as tourism, hospitality and agriculture, the country has no tradition in the field of technology and demonstrates little institutional or policy ambition to stimulate technology- and innovation-driven entrepreneurship.

The rest of the countries in the region—Albania, Bosnia and Herzegovina, Kosovo, Macedonia, Montenegro and Serbia—are all home to vibrant, albeit smaller and younger startup communities. Some of them, such as Serbia and Kosovo, have already produced some relatively significant internationally successful young companies, but in general they are less mature and have less experience. Overall, all startup communities across the region offer a welcoming and supportive setting for young people willing to explore entrepreneurship as a path. They play an essential role in encouraging the rise of an entrepreneurial culture within countries and across the region, and they are drivers of the many events and initiatives that raise awareness about entrepreneurship and that expose the current and future startup founders to current thinking, global best practices and inspiring examples. They also act as a support system for new entrants in the entrepreneurial arena who benefit from information, guidance and mentoring.

Many of the startups in the region are concentrated in the fields of technology and computer engineering given the historical strength of these industries. This has much to do with the fact that the region, and

Bulgaria and Romania in particular, have grown a competitive IT industry that has motivated a number of large multinational technology leaders to open local offices, which has created favourable conditions for the success of numerous small and medium-sized local companies.

Historical Background

It is not possible to make a comprehensive analysis of the factors that have led to the development of the current state of the startup ecosystem in Southeast Europe without tracing the roots of processes that go back to the region's Socialist period. The region is part of the so-called Eastern Bloc that included countries that were under Socialist rule between the mid-1940s and from 1989 to 1990. After the fall of the Berlin Wall in 1989, Socialist regimes dissolved and all countries in the region took a path of transitioning towards democratic rule and the development of market economies. This was a period of dynamic socio-economic processes throughout the region, with many painful transformations going hand in hand with the process of deconstructing Socialist systems and the rebuilding of democratic ones. Part of the region—most of the countries that belonged to the former Yugoslavia—experienced serious military conflict and the revival of religious and ethnic struggles. After strenuous international intervention, the region calmed down and began the peaceful process of accelerating its transition to a future aligned with the rest of Europe. In the mid-2000s, Slovenia (2004), Bulgaria (2007) and Romania (2007) become official members of the European Union, joining Greece, which was the first Balkan member in the EU, joining in 1981. Some of the other Balkan countries are on the path of becoming EU members, while others are preparing to begin negotiations.

One of the strongest features of the Socialist system in the Eastern European region was its education policy, with its strong emphasis on full access to basic education (elementary and intermediate) for the entire population, and access to advanced (tertiary) education for the greatest possible number. At the core of this policy is the belief that in a Socialist society, education is 'condition and factor in a country's economic growth, and the development of the individual which is determined by his education, becomes the source of personal satisfaction and the socio-cultural development of the nation' (Kluczynski, 1980, p. 2). As education enables the individual to achieve professional and social goals, it becomes a vehicle for building a developed Socialist society that is aligned with the scientific-technical revolution, which is fundamental to a nation's growth and development (Kluczynski, 1980). In addition to educating the population and securing widespread social benefits for children (Lokshin & Fong, 2000), Socialist governments provided guaranteed access to education, as part of childcare policies that enabled human capital development through wider access to economic activities for women (Meurs & Giddings, 2004). These policies ultimately resulted in higher rates of education compared to international standards at the time and a negligible level of illiteracy in most Socialist societies in Eastern Europe, including Balkan nations.

Part of the strategic development of Socialist Eastern Europe included their specialization in particular economic sectors following inter-industry trade patterns agreed upon among the member states of the Council on Mutual Economic Assistance (CMEA). Based on that, Bulgaria, for example, started to develop as a centre for mechanical engineering and electronics, while other countries specialized in machinery, consumer goods or extracting and supplying raw materials (Hallet, 1997). This meant significant and targeted public investment to build the social infrastructure to support such specialization, including the building of advanced expertise.

Widespread education and an emphasis on specializations in the engineering and technology domains contributed to the gradual building of advanced technical capabilities across several countries in Southeastern Europe. This involved the development of advanced technical education, scientific research capabilities and industrial enterprises employing large numbers of well-trained experts. These policies and investments lead to the nurturing of generations of advanced engineering and information technology experts. This technical expertise became one of the key factors in the formation of a professional community that gave birth to the first technical entrepreneurs after the fall of Socialism and the start of the transition to market economy status in 1989–90.

The (Re)Birth of Entrepreneurship on the Balkans

One of the more interesting questions relating to entrepreneurial dynamics in the region involves differentiating between today's entrepreneurs and those entrepreneurs who succeeded in the earlier period of the transition from a Socialist to market economy. Some of the more experienced members of the ecosystem, who witnessed both the earlier socio-economic dynamics and the developments in the regional entrepreneurial ecosystems after 2012, suggest that the true entrepreneurs in the region are those who succeeded in building successful businesses back in the earlier years of transition (1990s–2000s), when there was no economic infrastructure, when finance was limited and when societies were experiencing constant turbulence from the ongoing transformation. While public opinion in most countries is influenced by a suspicion that many of the successful business people from those years based their success on illegal or unethical practices that were possible in the murky waters of the times, there are also numerous success stories of individuals who managed to build internationally competitive companies based on strong business sense, entrepreneurial boldness and hard work. As one of the prominent members of the regional investor community, himself a serial entrepreneur, shared in an interview:

> *There is an older class of business entrepreneurs who started their business in 1990–1995 and managed to navigate crises, political changes and so on, and as far as I am concerned, they are the true entrepreneurs of this region. These are the people to whom we should be turning for advice and for guidance today.*

Many of these successful companies that were founded as entrepreneurial ventures in the 1990s and early 2000s are still around and are leaders in their sectors. While some of the larger ones operating in more visible sectors such as food and beverages, fast-moving consumer goods or entertainment are known to the wider public, there are many that are not (Peshev, 2017). These companies lack a public profile and are known only to a limited circle of employees and in their networks, but they represent success stories that provide a great deal of optimism about the potential competitiveness of enterprises created in these geographic locations.

For example, Walltopia is a market leader in the design and production of artificial climbing structures. The company was originally founded in 1998 by two young entrepreneurs and passionate climbers: Ivaylo Penchev and Metin Musov. Today, the company, headquartered in Sofia, operates on six continents with offices in the USA, UK, Canada, Germany, Austria, Russia and Asia Pacific and exports to more than 50 countries around the globe. It is a leader in the industry, not only in terms of market share and sales volumes but in innovation. The company invests heavily in sophisticated R&D facilities and operations that constantly explore new solutions in terms of material, design and applications. In 2018 the company employed over 600 employees around the world.

Another example of the resilience of the entrepreneurial spirit in the post-1989 period is IPS. IPS is a family-run company specializing in R&D and manufacturing of power electronic and energy conversion technologies. Through its specialized technology, the company is providing energy solutions in areas which have limited or no access to electricity networks, or where electricity networks are unreliable. Using cutting-edge innovative technology, it uses alternative sources to create intelligent off-the-grid stable energy solutions. Originally founded in 1989 in Bulgaria, the company remains headquartered there but operates across 56 countries.

In 2014, the company received a global innovation award in Germany for its Exeron system. In 2017, a Bulgarian engineering team led by the CEO and son of the original founder of IPS, Alexander Rangelov, developed the power solution for Elon Musk's Hyperloop capsule and received an innovation award presented by Musk himself.

In 2018, it launched its newest factory with a total floor area of 16,800 square metres, which will be the first 100% self-sufficient zero-emission manufacturing facility in Europe.

The publicity that this entrepreneur was given is the exception rather than the rule. In fact, the lack of publicity around the successful entrepreneurs of the past is a great missed opportunity. First, the lack of awareness of the many authentic stories of success of individuals who have achieved international recognition in a local context continues to fuel the general disbelief, especially among the older generation, that real and honest success is possible. Jakob Modéer, who has been working with different donor programmes across the region supporting the private sector and who currently heads the entrepreneurship programme of Swisscontact supporting startup communities in four Western Balkan countries, shares his observations about the lower profiles of successful people at the expense of the higher profiles of those who have succeeded for the wrong reasons:

> *In the Balkans, there is something called the 'businessman' that comes with a negative meaning. We all know these people made money easily, very quickly and in most cases through misuse of public funds.... These are not people who operate in a normal economy where price and quality matter... So, this image of easy money is still lingering on. Whether we want it or not, there are still forces that make heroes out of these people and of course they are not heroes. Where I come from, in Sweden, we have other heroes like Volvo, Scania and IKEA, and so people know them...*

The need to publicize the original success stories of older-generation entrepreneurs who succeeded by following ethical business practices is even more important given that some of them are identified as the triggers of current entrepreneurial processes. George Roth, who was born and raised in Romania but is now a successful serial entrepreneur in Silicon Valley and Honorary Consul of Romania in San Francisco, recalls:

> *I think the big turning point [in the entrepreneurship processes], at least for Romanians, was the acquisition of the first Romanian IT company by Microsoft in 2003. Radu Georgescu was the founder of the company called GECAD Ventures and the product they sold was an antivirus software called RAV. So this acquisition created the model for local entrepreneurs who saw how they could make money in IT. Until then they did not have that notion.*

Radu Georgescu started his entrepreneurship journey during the final year of his university studies. He started his company, GECAD, in 1992, and in 1994, he began working on RAV Antivirus along with three other software products. Three out of the four early products failed, but RAV software became one of the most important success stories in Romania's IT sector.

By 2003, the product was selling through a network of more than 150 resellers in 60 countries. Its popularity attracted Microsoft's attention, and in 2003 it bought RAV Antivirus from GECAD.

Radu Georgescu and the leadership team of GECAD refused to join Microsoft and remain in Romania as important catalysers of growth in the IT sector. They continue to develop new software programs including an email server, Axigen; an online payment platform, ePayment; an Electronic Software Distribution platform, Avangate; and a security solution (non-antivirus), Sentinet.

In 2010, the company was acquired by Allegro/Naspers becoming one of the successful exits in the region.

Somewhat surprisingly, the successful older-generation entrepreneurs are not celebrated and are somewhat disconnected from the new generation of entrepreneurs who seek their inspiration from role models from Silicon Valley rather than businesses that have thrived locally. This creates a certain degree of disconnect in the entrepreneurial value chain, limiting the opportunities for benefiting from locally relevant lessons learned and the adoption of established models of success. Jakob Modéer talks about this need to connect newer generations of entrepreneurs with the older generation of successful business people. According to him, bridging the generations would be one of the indicators of success for entrepreneurship in the region:

> The startup community here is very tech oriented and to some extent this is also because the philosophy of the lean startup and accelerator movement, which is influential here, comes from the US—Boulder, Colorado and Silicon Valley. We are trying to implement this thinking here and we have a different socio-cultural environment. I think the success of the [entrepreneurial] ecosystem in the Balkans will depend on its capacity to link to strong traditional sectors here. So, for Albania, it should be tech in tourism or tech in shoes; Serbia has a strong light industry and food processing industry, and so on. So, if we can apply the lean methodology and do hackathons with a more specialized focus, and try to involve the big companies so they can open up a little bit, that should be very interesting. So, I think this will be one indicator of success—how the ecosystem leaves its own bubble and takes the discussion to the older generation of business people, to the corporate guys.

In some sectors, notably IT, and countries, this continuity between generations of entrepreneurs has been easier to achieve.

The Growth of the IT Industry

As discussed earlier in this chapter, the development of strong computer engineering capabilities was part of the national strategy of Socialist Bulgaria and was supported by regional cooperation strategies in the Eastern Bloc. Continued investment in the sector and the effective nurturing of generations of experts produced recognizable results. As one of the pioneers of the contemporary information technology cluster commented: '*If you look at the World Bank statistics from the 1980s, you will note that Bulgaria was one of the biggest players in the global trade for computer components after the US and Taiwan.*'

This advanced capability was the foundation on which today's regionally significant innovation ecosystem was built. Educational traditions, advanced skills and sophisticated expertise in computer technology were recognized as the right ingredients for the emerging global competitiveness of the sector. While the turbulent 1990s saw some success stories, with the launch of local IT companies offering products and services to local and international markets, the real consolidation processes in the IT industry in Bulgaria began in the early 2000s. At the time, Bulgaria, Romania and Slovenia were on track to become EU members, which fuelled optimism and channelled important resources for economic development from pre-accession European funds as well as other active donor programmes, including the United States Agency for International Development (USAID). Among the active donor programmes in Bulgaria was the USAID-funded Bulgarian Enterprise Growth and Investment Project (EGIP), focused on strengthening the competitiveness of several high-potential economic sectors, including ICT. Designed around Michael Porter's theory of competitive advantage, EGIP's support for the ICT sector includes awareness raising and education of key stakeholders on competitiveness principles as well as targeted facilitation of dialogue and collaboration within the sector (Silcox, Lobron, & Nathanson, 2005). Even though the donor investment in the ICT sector channelled through EGIP is miniscule compared to the funds invested in other economic sectors during the same years by the EU and other donors, the impact was profound. The early 2000s saw the formation of an ICT cluster that included important processes such as industry collaboration, targeted policy changes and foreign investment promotion. During this period, some of the global industry leaders such as HP and Microsoft already present in the country established fast-growing local offices. A number of locally established companies also developed quickly and achieved success on global markets. By the mid-2000s, after Slovenia had become the first new EU member from the region and Bulgaria and Romania were gearing towards joining the Union, it became clear that the newly established JEREMIE (Joint European Resources for Micro to Medium Enterprises) fund would channel significant financial resources to EU member states for small and medium-sized enterprises by means of equity, loans or guarantees (EIF, 2015). Although public institutions and government agencies were not particularly supportive at the time, the ICT cluster was actively led by the private sector and began to prepare for the opportunities afforded by the JEREMIE fund.

An interesting debate in the analysis of the history leading to the formation of a startup ecosystem in the region is whether public institutions and government agencies have contributed in any way to the success. Many members of the regional entrepreneurial community, especially startup founders and younger entrepreneurs, are sceptical about the role of government and believe that the best a government can do is to stay away from the ecosystem. Their explanation for the success of startup processes in Southeast Europe is that public institutions somehow ignored what was happening in the IT entrepreneurship arena. More experienced members of the ecosystem, including serial entrepreneurs, senior investors, company executives and professionals with experience in the public sector, claim the opposite and express the opinion that the public sector is essential to the success of any ecosystem. In fact, they suggest that the main reason that countries like Bulgaria, Croatia, Romania and Slovenia have thriving entrepreneurial ecosystems today is that at different times over the years between 2000 and 2015 their governments included leaders with an affinity for entrepreneurship and a good understanding of the factors that are needed to encourage innovation and new venture formation. The lack of such a political vision in Greece, for example, is identified as one of the main reasons why the country has not managed to create an effective entrepreneurial IT community despite its ability to access the same support mechanisms and funding as the rest of the EU members in the region. In that sense, even if government-level support in Bulgaria, Croatia, Romania and Slovenia has been inconsistent and disrupted by political dynamics or frequent government changes, it may have been essential for their success. A former senior public official from one of the countries shared the view that some of the most important developments in the startup landscape throughout the region have happened because certain ambitious politicians have at some point taken on them as their 'pet projects' that they have stubbornly pushed through alongside the main responsibilities and priorities required by their political programmes. In many cases, these were younger-generation politicians with a foreign education and an advanced understanding of the principles and structures that can unlock innovation processes in economies. Some of them acknowledged that the lion's share of their efforts went into managing internal opposition from government colleagues who did not understand or see the importance of supporting an entrepreneurial ecosystem. Among the key decisions that were made by such political leaders were to:

- Attract and channel EU funds towards innovation and startup processes in the respective countries
- Facilitate the design of less cumbersome and more efficient funding mechanisms for startups and young entrepreneurial ventures to eliminate heavy bureaucratic intervention by government institutions
- Direct investment towards physical infrastructure that supports innovation and entrepreneurial activities
- Facilitate coordination between public and private sector stakeholders to achieve more through consolidated efforts
- Support private sector-led initiatives in the area of innovation and entrepreneurship

One of the key milestones in the development of the regional startup ecosystem was the launch of two early-stage investment funds in Bulgaria:

Eleven and *LAUNCHub*. These are the intermediary mechanisms created to channel €21 million from JEREMIE towards entrepreneurship acceleration and seed financing. The primary targets of the funds were innovative businesses and technology companies. Even though they were located in Bulgaria's capital city, the funds attracted beneficiaries from across the region. The launch of the two funds has fuelled a transformation in Bulgaria's startup and entrepreneurial scene and has accelerated the country's position as the regional centre of entrepreneurial activity. During the years following the launch of the two funds, more than 500 new ventures were created, attracting over €200 million funding in addition to the €21 million provided by JEREMIE. The estimated total volume of startup exits surpassed €500 million for the years up to 2017 (Move.bg, 2017).

The availability of funding is the next key ingredient for building success in combination with other positive factors such as a thriving and well-organized ICT industry, a well-educated and highly skilled workforce, an overall stable socio-economic environment and supportive institutions. In addition to channelling investments towards new ventures, the two funds consolidate other processes in the ecosystem and stimulate its regionalization. Sofia has become a centre for numerous startup events and initiatives that attract leaders in global innovation and the startup scene to the region. Entrepreneurs and business professionals from across the region have the opportunity to network with each other and with big names from leading innovation companies and global startup hubs. It has also become a preferred setting for the launch of many associated initiatives led by the corporate social responsibility programmes of large corporations or by non-government organizations, educational institutions or even private individuals. During this period, the idea to create Sofia Tech Park—a modern space that will offer physical infrastructure for innovation, bringing together all the important players in the startup ecosystem, including investors, entrepreneurs, research entities, educational institutions and other supporting organizations in one place—emerged. The official opening of Sofia Tech Park took place in 2015, becoming the first science and technology park in the region, established with the ambition of serving as the next-level platform for accelerated innovation through the exchange of knowledge and ideas between academia, business, government and society on a regional level. The scientific research facilities and physical infrastructure covers an area of 40,000 square metres and includes applied research laboratories, a general incubator, an innovative lecture and education space, areas for new technology demonstration and pleasant green spaces.

The Development of a Young Ecosystem

This section looks in more detail at how each of the six elements of the entrepreneurial ecosystem introduced in Chap. 1 has developed and shaped the current startup landscape in the region. We will look at entrepreneurial finance and how its availability in some countries in the region transformed them into regional centres, as well as how the limited availability of finance for the more mature phases of startup growth has created some challenges for local entrepreneurs. The availability of government programmes and support agencies will be discussed to reveal how the affinity for entrepreneurship of some individual politicians has played a key role in creating a favourable setting for the growth of a regional ecosystem. The important role of academic and research institutions will be discussed in light of the fact that many of the successful entrepreneurs in the region have been shaped by the education they have received, but we will see that educational institutions are still catching up with their role in the ecosystem. One of the most interesting considerations in the context of this book is how the market context and private sector sophistication of a region that has undergone dynamic transition processes supports or impedes the success and competitiveness of local entrepreneurial ventures. The development and quality of physical, legal and human infrastructure will also be analysed to understand how advantages and disadvantages in this domain intervene with the entrepreneurial dynamic. Last but not least, this section will zoom into regional culture and mindset specificities that contribute to the challenging and at times unusual context in which the venture creation process unfolds.

Entrepreneurial Finance

As discussed earlier in this book, one of the key elements for a successful entrepreneurial ecosystem is access to finance. With the growing importance of entrepreneurial activities in the region, the availability of capital has become a significant factor. Whether this is the main factor in the success of startup and entrepreneurial activities throughout the region is a question that triggers different opinions across the region. Many entrepreneurs consider the availability of capital as the main factor in the changing entrepreneurship dynamics of the region and think that the launch of targeted funding for new ventures through the European Union was the key milestone. Entrepreneurial finance made available through the two venture funds in Bulgaria—Eleven and LAUNCHub—is frequently identified as the reason for the country becoming a regional startup and entrepreneurship centre. Deeper analysis of the processes that accompanied the availability of venture capital suggests that the availability of finance played a catalysing role thanks to the favourable environment which included other important factors that were already in place. The fact that some of the other countries in the region had access to similar financial resources, even if on a smaller scale, but still failed to achieve the vibrancy of entrepreneurial processes observed in Bulgaria, supports this thinking.

Finance for the Full Growth Cycle

One of the weaker aspects of the availability of capital in the region is the limited opportunity for accessing funding for growth beyond the seed stage. Eleven and LAUNCHub manage to fuel support for many early-stage companies but do not have the mechanisms and funding to service startups in later growth stages. Some members of the community even state that many otherwise successful ventures that were recipients of seed investment failed to thrive due to a lack of follow-on funding to support them beyond the initial stages. Kiril Petkov, founder of one of the successful Bulgarian startups and an important influencer in the ecosystem, notes:

> Things in the region, and in Bulgaria, have been developing very well. Just recently I was travelling and in an article about entrepreneurship in the airline magazine I noticed that Sofia starts appearing as a startup hub. We were very lucky that there were European Union funds that were invested in this field and the two accelerators Eleven and Launchub were established to support relatively early ventures. This made this place a good place with good access to capital for very early-stage companies... Secondary stage funding (up to several million) is where the hardship is now. We see some progress there with funds like Blackpeak but this is still relatively limited with the existing local capital options.

While the initial design of financing mechanisms launched for the region through the JEREMIE fund intended to secure funding for the different stages of startup growth, complicated negotiations and political dynamics made later-stage funding less readily available. Some mechanisms were available, including through selected banks, but overall the opportunities were much more restricted than for seed stages. According to Momchil Vasilev from Endeavour Bulgaria, access to capital is a common problem for ecosystems like those in this region:

> One common problem for small markets like us is the lack of sufficient capital for all stages of development of the enterprises. In some markets in the region, like Bulgaria, some of the phases are covered but others, like the countries from former Yugoslavia, most of these are not there. This is the fuel that entrepreneurship runs on. This is not to say that entrepreneurship cannot survive without that and on its own but scaling and speed of entrepreneurship is vastly accelerated through the presence of sufficient capital. This of course, has something to do with how attractive these markets are for investors and part of our organization's mission is that.

Local or Foreign Investment Sources

A key point about the availability of finance in the region is the importance of having well-functioning and effective local and regional investment funds. Local and regional investors are familiar with the socio-economic context and can evaluate investment opportunities more effectively. They are also much more likely to be engaged and support ventures with expertise, mentoring and networking than external players. They are also more interested in the relatively small investments that are in demand from new ventures while foreign funds are looking for much larger opportunities (tickets of €5–10 million). New ventures that are launched and operate in this region fall well below this size, which is why foreign investors are not the only solution. Mircea Pasoi, a Romanian-born entrepreneur who is now based in Silicon Valley, shares his observations about the potential interest of US investors in Romania and the region:

> I have seen little interest in US-based investors to scout for opportunities in Eastern Europe. Even here in the US, their focus is concentrated mostly in San Francisco and New York. Yesterday I was in conversation with one investor and trying to convince them to go to a tech conference in Romania, which is very attractive and successful in the region. Even though the organizers will cover all their travel expenses, they still did not want to go. We are still seen as an area for cost optimization that offers good talent at low salaries but I would not say that the thinking of investors here is that people in the region have interesting ideas.

Access to finance by new ventures in the region is key because it also triggers other activities that typically accompany the growth of an ecosystem. Startup events initiated or supported by the two funds catalyse a lot of valuable processes, such as attracting relevant expertise and mentorship to the region; the exposure of local entrepreneurs to global role models, as well as their stories of success and failure; the advancement of entrepreneurial culture and skills; the rise of many allied initiatives and services for the entrepreneurial community; and the formation of an angel investor community.

Government Programmes and the Public Sector

Even though the main source of energy fuelling the rise of entrepreneurship in the region comes from skilful and energetic private initiatives, public policies and government agencies have a role to play. There are numerous ecosystem players who believe that the best way for government to support entrepreneurial activity is to stay out of it. The lessons of the region, however, suggest that a successful ecosystem cannot be nurtured if public institutions are completely isolated from the process. The opinion of more experienced stakeholders and entrepreneurship experts is that government should be involved as a hand-off, arms-length investor in the ecosystem, specifically in areas where the private sector is not brave enough to provide capital. Momchil Vasilev, Corporate Finance expert and Managing Director of Endeavour Bulgaria, notes:

> Government support is something that entrepreneurs in their true nature are not accustomed to and do not seek. But where the government can be very supportive is in creating the infrastructure, creating the environment where entrepreneurs can develop. I am not talking about any subsidies or other forms of support that may alter the market in a negative way but about creating a solid and stable environment where entrepreneurs can look forward and plan with certainty.

While in some countries funding for early-stage ventures is channelled through public institutions, the more effective mechanism used in the Balkan region is to put the financing operation in the hands of the private sector. This makes it possible to overcome the typical slowness and bureaucracy of governmental departments, and to ensure that investment opportunities are evaluated from a market perspective. In addition, private fund managers usually provide ventures with much more than capital—they secure mentorship, relevant expertise and day-to-day support that facilitate market success. These support services are essential for the success of entrepreneurial ventures, but government agencies and public servants do not have the experience and skills to offer these services in any meaningful way.

Another significant benefit of leaving the investment role in the hands of the private sector is that it motivates private investors to build on available funding with their own resources. The launch of the two accelerator funds in Bulgaria, for example, created a favourable setting for the appearance of many angel investors and motivated the addition of private capital to the funds made available through EU and government sources.

Jakob Modéer from the Entrepreneurship Programme of Swiss Contact talks about the dialogue between entrepreneurs and the public sector:

> [The] private sector here has to grow up and begin taking care of itself. It has to stop looking to the public sector to fix their business… [An] enabling environment is of course needed but we need to have here a private sector that can stand up to government and say: 'Look, for this to work, we have to make a deal. We have to produce [a product] and we have to generate revenue, and then you can tax me. Do not tax me before I make a profit.' [The] private sector should be able to engage government in dialogue and express the exact demands that will create a favourable setting, not just wait for help.

The area where government can play an essential role is in creating an appropriate environment for entrepreneurship. Given that talent is one of the essential elements needed for a thriving entrepreneurship ecosystem, the public sector has a role to play in encouraging talented young people to stay in the region or even return from living abroad. The formation of a regional ecosystem with available financing and support services has already triggered such an initial process in some countries. More can be done to ensure that when people return they find the necessary conditions for a good standard of living.

Policy initiatives are also an important area in which public institutions can enable or slow down entrepreneurial processes. Targeted policy decisions can create conditions that make it attractive to start new ventures, and to invest time and resources in startup projects. Ease of registration and legal procedures around the launch of a company is also something that can be secured with relevant regulatory steps. Policies and regulations are also key when it comes to making capital available and easy to access.

Lessons learned from the region suggest that government has an important role to play in consolidating resources and directing them towards the development of physical environments and infrastructure in ways that can support the entrepreneurial dynamic. Evgeny Angelov, Chairman of the Bulgarian Private Equity and Venture Capital Association (BVCA), commented:

> In all of the places where [innovation] works well, there is significant clustering of the right people in one place: different startups mixing with different large companies. Creating a physical environment to enable innovation is one of the ways in which a government can support innovation processes.

In the case of Sofia Tech Park, the decision made by Bulgarian government officials that allowed for this to happen was to consolidate available funding, originally allocated for multiple small innovation centres throughout the country, and to execute a single large-scale project in Sofia to facilitate collaboration between more stakeholders, thereby triggering regional impacts. The investment initiated by the government served as a foundation for the engagement of private sector players, and in 2018 Sofia Tech Park is home not only to different startups but also to some of the largest leaders in innovation such as SAP, Microsoft and Walltopia.

Weaknesses and Challenges with Governments as Ecosystem Members

One of the biggest challenges associated with the role of public sector in stimulating entrepreneurship is that the types of initiatives that naturally fall under a government's portfolio require time. At the same time, especially in a rapidly developing region where countries experience frequent government changes and politics can be turbulent, continuity in projects is a challenge. This makes politicians hesitant to initiate impactful initiatives such as reforming and creating educational or research structures to nurture an innovation culture, developing large infrastructure projects that create the right setting for entrepreneurship, supporting long-term mechanisms for venture funding and so on. This is why even some of the most effective government interventions that have been executed in the region have not followed a straight path. As an interviewee from the public sector noted, in some cases, effective interventions are the result of a combination of the right people coming together and some lucky circumstances that enable them to make decisions that buck the political odds.

Academic and Research Institutions (and R&D Transfer)

As discussed earlier in this chapter, wide access to education and educational traditions in specific domains such as computer technology and engineering during the Socialist period of the region are among the key factors that constitute the foundation for today's entrepreneurial dynamics. Education and knowledge creation are essential for any ecosystem, but embedding them takes a lot of time. The high quality of the technical education found in the social systems of the past is still yielding results today and has served as a basis for the enabling of a competitive ITC sector in Bulgaria and the region.

Educational institutions are important players in any entrepreneurial ecosystem, not only because they can shape the skills that future entrepreneurs need but also because of the networking and community effects they provide. Many of the successful Bulgarian tech startups were founded by college friends and were initially launched in student dormitories. In Macedonia, almost the entire tech entrepreneurship community is linked to a graduate programme in digital communications that was run for a few years at one of the local universities. The programme no longer exists, but during the few years in which it was active, it attracted and trained many of today's startup founders or staff members of organizations providing services to entrepreneurs.

The Lack of 'What Is Next' Thinking

Unfortunately, during the many years of transition, the educational sector in the region remained slow to adapt to contemporary realities, and especially to the need to be linked to entrepreneurial processes in the region's economies. While Bulgaria, Romania, Croatia and Slovenia are home to some competitive higher-education institutions that offer some relevant content and align themselves with current thinking in education, many of the traditional universities throughout the region have done little to reform their systems or content. A lack of funding and of contemporary vision have made them unattractive to the ambitious and the smart so many young people are leaving the region to seek better education in Western Europe or North America. While this is bad news for local educational institutions, it is not necessarily bad news for entrepreneurial dynamics in the region. Some of the countries with more advanced startup processes are beginning to see young people return to the region to start their own ventures. In some cases, they come back with international college friends to initiate joint projects. If entrepreneurial dynamics really manage to unlock a process like this, startup ecosystems may offer a solution to one of the most dramatic socio-economic problems of the region—the brain drain. Mira Krusteff, co-founder of GEM Bulgaria and entrepreneur in education, comments:

> The most recognizable entrepreneurial culture and skills are observed among people who have studied and lived abroad, and then have come back, or they have travelled or worked for a foreign company, or a family member has been abroad. There is always a link with 'abroad' or with being able to use a foreign language.

In fact, the gap between entrepreneurial processes and tertiary education institutions seems to be growing wider. Recent analyses reveal that fewer companies and startups in Bulgaria now rely on universities as a channel for securing new talent (Move.bg, 2017). The fact that many of the large companies, especially in the ICT sector, have created their own internal academies to develop the skills that they need is further indication that formal academic institutions are no longer providing the solid preparation that is needed for competitiveness and growth in this important sector. While technical education has been a source of advantage for decades, rapid developments in this domain require systematic upgrades of curricula and learning techniques. As the serial entrepreneur, mentor and regional startup expert Max Gurvits notes in an interview:

> When you look at the way Silicon Valley has transformed over recent years, you see a very clear shift from easy tech to hard tech so almost all applications [in accelerators] now are about robotics, virtual reality, everything linked to big data and artificial intelligence, so everything in technology is getting much more complicated and sophisticated… I have the feeling that from a purely technical point of view, we are beginning to lag behind

> because it is known that educational institutions in the region are very good at the fundamentals such as mathematics but they do not really offer advanced tech skills. I fear that there is a risk that because of this we might lose our technological advantage.

The challenge for educational and research capabilities is that their development takes time. Even though some individual initiatives linking universities to some of the entrepreneurial endeavours in the region exist, there has been no consistent and strategic effort to align education with innovation processes and needs. An encouraging aspect observed across universities in the region is that many have included subjects such as entrepreneurship and new venture creation in their curricula.

Entrepreneurial Education at Early Stages

While the situation in tertiary-level education is a bit pessimistic, there have been some active developments at primary and secondary education level. The persistent efforts of programmes such as Junior Achievement have succeeded in adding entrepreneurship as a required subject in the school curriculum in Bulgaria. This ensures that students will be exposed to the principles and models of entrepreneurship at an early age. Experts Milena Stoycheva and Petko Rusev from Junior Achievement-Bulgaria, which pioneered the link between entrepreneurship and education in the region, share their thoughts:

> After a lot of work in 2016 on the new law on education, there has been a gradual introduction of entrepreneurship as a required subject in the curriculum in primary and secondary education. Part of this involves the introduction of a concentration in Entrepreneurship for middle school, which allows students to choose this as the special focus of their studies… The big problem is the education and training of teachers, because these topics cannot be taught using the same approach these teachers use in other subjects. This requires a lot of time and effort… Another big problem is that actually the big universities are hesitant to participate in these initiatives so where we are able to make some progress is in small universities. All of this leads to a very slow pace of change in the education sector… This lack of investment in education leads to a big problem for the future of the

> ecosystem as there are many activities involving the current participants in the entrepreneurial community but there is not much that is being done to generate more and grow this community further… This is why one of the most important areas of work for us is working with startups in universities who learn how to experiment and how to fail.

Aligned with this is the rise of many private sector-driven programmes that train children in basic skills in the area of coding, computer design and robotics. Delivered in different formats, such as extracurricular programmes, weekend courses or summer camps, they offer an alternative opportunity for nurturing the next generation of entrepreneurs with the right thinking and the right skills.

Marketing and Business Culture

While those who are familiar with entrepreneurship in the region acknowledge the high quality of its technical talent, one of the most frequently shared observations made by foreign and regional experts relates to the weak business and marketing skills of local entrepreneurs. This is largely due to the lack of a tradition in business education and the region's slow pace in adapting educational content to compensate for that. At the time of writing, most of the countries in the region have access to internationally competitive Master of Business Administration (MBA) or executive education programmes in the area of business. Their primary target is mid- to high-level management of larger companies. However, because few entrepreneurs go through these programmes and most emerge from technical academic courses, the gap in business capabilities has not yet been plugged. Although discussions have taken place about linking business education programmes for entrepreneurs with accelerators and other funding mechanisms, this has not been carried through. Some regional experts think that many of the failed startups over the past years have developed good products but have failed to identify the right market or to adopt the right go-to-market approach. Larry Biehl, philanthropist and founder of the Bulgarian Entrepreneurship Centre, highlights the importance of business skills and identifies the lack thereof as the main disadvantage for entrepreneurs in the region:

> Business building skills are important because they allow you to understand the evolutionary aspect of starting, building, scaling and selling the business… What is also missing are sales and marketing skills… Where an entrepreneur needs to start is by thinking about who are the potential users for this service or product, who are your competitors, what is the level of user satisfaction you need to deliver. These are the kinds of questions you need to be able to answer. So, sales and marketing skills, and business building skills are what marks the difference between entrepreneurs here and entrepreneurs in the United States.

Market Functioning and Private Sector Sophistication

Among the factors identified as favourable for entrepreneurial dynamics is the overall positive socio-economic context in the region, especially in countries that are part of the European Union. Some of the economies in the region, such as Bulgaria and Romania, demonstrated stable growth and experienced little decline even during the years following the economic downturn after 2008. Iskren Krusteff, founder of GEM Bulgaria, observes:

> The environment could be much more favourable but if you look at Africa and various countries on that continent, you see some of the highest rates of entrepreneurship. It is true that entrepreneurship there is necessity-driven because people have to survive but we also have that mindset of survival. So, the environment is favourable enough to nurture entrepreneurship in the region.

Some of the support, including donor funding, that has been directed at entrepreneurship is also producing a positive impact. The head of the Entrepreneurship programme of Swisscontact in the Western Balkan countries, Jakob Modéer, comments:

> We see growth overall, and there are more and more people involved in the ecosystems now than two years ago. There are more programmes, there are

> more organizations that are moving into this space, there is more private money than there was two years ago, and this is because there are more competitive IT companies offering outsourcing services. They are becoming the 'mother ship' that is a force that is driving the acceleration, which I think is a very important development. I am curious to see how their initiatives will play out in comparison with the ones that were set up by donors.

While economic growth has been positive, compared to some of the places with the most vibrant entrepreneurial processes, the region's economies are still young and less sophisticated. This is both good and bad news. It is bad news as it means that private sector players as well as policy-makers and non-government stakeholders have less experience of behaviours and tools that stimulate innovation and adaptation to economic realities. At the same time, it is good news, as the lack of advanced systems and tools represents a real opportunity to innovate by adapting solutions that already exist and work in other more advanced economies to the local context. One example of such a success story is the Kosovo-based Gjirafa—a Google-like content and e-commerce platform offered in Albanian that dominates in the region and in the significant Albanian-speaking diaspora in Europe. The success of the company, which is one of the startups to have attracted the highest level of investment across the Western Balkans, is based on adapting services and solutions that exist in more advanced economies to the local context and needs. Another similar success story is *Tickey*, a Bulgarian startup that offers digital services optimizing the public transportation experience to increase its attractiveness and to discourage the use of cars in large cities. The digital services that are part of the platform are not revolutionary and exist in some advanced economies with more sophisticated public transportation systems but represent an economic opportunity and generate important social impact in countries in the region.

This raises an interesting question about whether startups in the region would have more of an impact if they produced more ventures on a regional scale and focused on providing solutions relevant to the region, or whether they should aim to compete in the global arena. Milos Dosen, a Serbian entrepreneur and investment-readiness expert, comments:

> I think that in general you can always establish a good local business with good profit levels, and do quite well in local markets. Not everybody has to scale up and go global. It is better to offer some good local service or product, and to make it happen here rather than to lose everything for the sake of being a global company. There is enough opportunity here.

From a regional perspective, more locally focused entrepreneurial activity and innovation will have a profound effect on the socio-economic environment and will accelerate the closing of the gap between the region and more advanced economies in Western Europe. Startups with a regional focus are also likelier to remain in the region and create local jobs and economic opportunities. Global success stories, however, will also have a positive impact. While many of the local startups that pursue success in Western Europe and North America eventually move some part of their operations closer to their target markets, they still leave part of their operations in the home country. Big success stories have an important reputational effect and can really accelerate the visibility of the region as a significant centre of entrepreneurial activity and innovation. This is how the success story of Skype put the Baltics on the global ventures map.

Mircea Pasoi, a Romanian entrepreneur who is now based in Silicon Valley but began his entrepreneurial projects back in Romania, explains that one of the key factors that made the IT sector so important in the region were the early successes, which served as models:

> Some of the first success stories, which inspired entrepreneurship in Romania, were in the tech domain and that is an important reason why this sector picked up. In the early 2000s in Romania, we saw some cases of startups that grew fast and then were acquired by Microsoft and by Adobe… They were very inspiring to me and to others like me.

While some sectors of the economy are characterized by a lower level of sophistication, the competitiveness of the ITC sector and the fact that it has successfully plugged itself into the global marketplace have contributed to its development and sophistication. Many locally established ITC companies service global leaders and interact with the innovation drivers in the sector, which inevitably influences their strategies and capabilities. George Roth, entrepreneur and Honorary Consul of Romania in San

Francisco, shares his perspective on why he thinks the IT sector picked up so well:

> In most countries in the region, growth in IT started as an outsourcing business where success came based on local talent, good education and cost effectiveness… To some extent, the countries also realized that exporting IT services is easier than exporting material goods because you bypass a lot of complexities, including corruption. After all, you do not need the same level of approvals for virtual services as for material goods. So, in the Balkans the IT sector was successful because of the great talent pool with a good level of education and because IT businesses were easier to start than other businesses in this environment.

The fact that some of the largest companies in digital solutions and computer technologies have large offices in the region also contributes to the sophistication of the sector. Companies like HP, Microsoft, SAP, VMware and Experian have set up large local offices that contribute to the global product and service value chains rather than simply providing basic back-end or maintenance services. In addition to being a springboard for a lot of ICT professionals, some of whom join the startup community after gaining corporate experience, these companies also contribute to the local innovation dynamic. Many of them initiate and support hackathons, innovation competitions and startup challenges for students or for young entrepreneurs. Some have initiated collaborations with universities and have invested in the development of academic programmes that provide the knowledge and skills that are relevant to their cutting-edge industries.

Limited Market Size

One of the particular challenges of the market context in the region is the size of local markets. Most of the countries in the region have relatively small populations and limited market size (see Table 3.1). The lower purchasing power of the populations compared to more advanced markets is also considered to be a barrier. More optimistic regional experts, however, highlight the fact that the smaller market size should be looked at as a

Table 3.1 Economies in the Balkans (in US$)

Country	EU membership	Population	GDP (purchasing power parity)	GDP per capita (PPP)
Albania	*Accession negotiations not started*	3 million	$35.9 billion	$12,500
Bulgaria	*EU member*	7.1 million	$152.4 billion	$21,700
Bosnia and Herzegovina	*Accession negotiations not started*	3.9 million	$43.9 billion	$11,400
Croatia	*EU member*	4.3 million	$100.2 billion	$24,100
Greece	*EU member*	10.8 million	$299.5 billion	$27,800
Kosovo	*Accession negotiations not started*	1.9 million	$19.4 billion	$10,400
Macedonia	*Accession negotiations not started*	2.1 million	$31.6 billion	$15,200
Montenegro	*In accession negotiations*	0.6 million	$10.9 billion	$17,400
Romania	*EU member*	21.5 million	$474 billion	$24,000
Serbia	*Accession negotiations not started*	7.1 million	$106.6 billion	$15,200
Slovenia	*EU member*	1.9 million	$70.36 billion	$34,100

Source: Based on 2017 data published in the CIA Factbook (CIA Factbook, n.d.)

challenge but also an opportunity. If the local markets are small, then companies should aim from the start to compete in larger foreign markets, where they can cut their teeth more rapidly. According to Marius Ursache, an experienced serial entrepreneur from Romania:

> Market size is a problem and a blessing at the same time. Ideally you should start your business in your beachhead market, a place of easy access where you have connections and you understand the culture, and usually these are our countries. This is an advantage because you can focus on the local market, and if you create a successful product or service which is validated in your market, you can expand to other markets.

The smaller size of the home markets is another reason to look at the opportunities in this region through the prism of regional integration.

Although the individual markets in each country are limited to just a few million potential customers, the entire region is populated by almost 65 million people, which translates to much bigger potential. This is why one of the topics most frequently raised by experts in entrepreneurial circles is regional cooperation and linkages between the ecosystems in the different countries.

Another potential way of overcoming the limitation of market size is to focus innovation and new venture creation processes on niche and highly specialized offerings rather than on scale. This can lead to the generation of similar profitability as large-scale businesses with a logic based on differentiation rather than volume. In the words of Momchil Vasilev from Endeavour Bulgaria:

> Our geography is often considered to be a problem and a handicap to entrepreneurship but I also think that is quite an advantage. Entrepreneurs coming from geographies which consist of small and underdeveloped markets deal from day 1 with challenges that make them stronger; from day 1 they seek to reach decent scale and thus have to think about expansion in different countries, which means dealing with different languages, different cultures, different systems…these are not easy environments to work in. So rather than starting a business in a large market where you can harvest the opportunities that are present and exist already, here you have to make the opportunities. That makes indigenous entrepreneurs far more resilient and capable of surviving in dynamic and changing environments. I was recently joking in a conversation that if you start a business, and make it in the neighbouring countries across the region, then Germany becomes just the next market you go to. On the other hand, if you start a business in Germany, coming to neighbouring Poland could be a shock. This is why I think that our geography could be an advantage.

Infrastructure

The advancement of the ITC industry in the region, including the presence of large multinational players, has contributed to the development of a relatively strong infrastructure supporting the use of digital technologies. Bulgaria, Romania and Slovenia, for example, which are home to the greatest number of successful new ITC ventures, are amongst the countries with the fastest internet connectivity, with speeds that are comparable to countries with advanced technology sectors and startup ecosystems ('Fastmetrics: Internet Speeds by Country—Fastest Internet In The World Map,' n.d.).

Another important aspect of the infrastructure supporting entrepreneurial dynamics are co-working spaces and physical infrastructure that serve as a setting for collaboration and innovation. This includes the previously mentioned Sofia Tech Park, a physical space created with the purpose of bringing together as many players from the regional ecosystem as possible. The main development, however, is the rise in the number of co-working spaces across countries in the region, especially those with more advanced startup communities. In Bulgaria, Romania, Slovenia and Croatia, co-working spaces have been appearing and they are growing in number, becoming a setting for regular meetings, events and competitions. They have become an attractive alternative to traditional business centres, even for some older and well-established companies seeking to maintain a culture of innovation and entrepreneurial thinking and actively take part in the regional entrepreneurial communities.

Whilst the region has a solid basis in terms of directly relevant infrastructure, the legal and commercial context poses some challenges associated with weaknesses in the commercial environment and ineffective legal systems. Some countries in the region offer attractive corporate tax rates, but others could definitely do more to offer a more favourable setting for existing and younger businesses.

Many of the countries in the region have a reputation for suffering from corruption and crime. There is an interesting ongoing discussion across the regional ecosystem about whether corruption, real or perceived, impedes entrepreneurship. Many founders, especially of startups that do not interact much with public institutions or rely on public services, consider corruption to be an inherent evil of the system but one that does not really influence their prospects for success. Marius Ursache, an entrepreneur from Romania, notes: 'Nowadays, corruption is not as bad as it used to be. In terms of daily operations for businesses, I think it is not an issue, unless your client is the government or some big organization.' Several startup founders in Albania stated that:

'Corruption affects some markets, but not necessarily innovation. Corruption represents costs and usually in startups you don't have to face these costs.'

Other members of the community, especially people who interact with the government and work within the ecosystem at a higher level, suggested that corruption and institutional inefficiencies across the region should not be ignored and that, sooner or later, they would begin to impede growth and innovation. In many cases, the problem is associated with the fact that corruption makes institutions slow and ineffective, making them unreliable partners and unable to play their role in supporting development processes in the ecosystem. Evgeny Angelov, Chairman of the Bulgarian Private Equity and Venture Capital Association (BVCA), notes:

> The biggest problem [in the region] is rule of law… If you think about it, if you have 10% corporate tax rate, low energy costs, every company in Europe should be based here. But the fact is they are not because of hidden taxation, uncertainty around ownership of property, etc.… Corruption certainly has a big impact. I wouldn't say that it is the biggest problem but there is the whole burden of the bureaucratic machine of the government. There are enormous inefficiencies in the system… this is the biggest problem because it wastes a lot of time and breeds low-level corruption. Luckily, most of the entrepreneurial companies that are involved with technology have very little interaction with the government, so I don't think they are really exposed to it.

In a conversation about the history of the US support for Bulgarian entrepreneurs, Nancy Schiller, CEO of America for Bulgaria Foundation, recalls the many companies that were supported back in the 1990s:

> There are many companies that we helped, which are still thriving in the country today but if we looked at our real success, it is working with Bulgarians—whether they were entrepreneurs or staff—working together to build this success. So, it is not individual companies or groups of companies that are the main achievement, it is those people that were behind them and have worked with us, and have afterwards blossomed in different career paths. That is the real success story.

One of the pillars of the success of entrepreneurial processes in the region is human capital. Smart and well-educated people with advanced technology skills can command salaries that are high for the region but very competitive in comparison to Western Europe and North America. Easy access to education in the rest of the European Union has made it possible for many young people in Bulgaria, Croatia, Romania and Slovenia to develop competitive profiles by studying abroad, returning to earn salaries that secure a very good lifestyle at home, whilst being lower than those offered in Western countries.

That said, there are two main challenges associated with the human capital advantage which, as we have seen, has fuelled the growth of a successful ICT industry in particular and, to some extent, the entrepreneurial communities more generally. The first is that the majority of young people, especially those with ambition and talent, still find it preferable to remain in the West after completing their studies. Many of them start successful careers and find few incentives to return home. This is especially relevant in the professions, which are difficult to practise at home due to the lack of development in the field. For example, for young people who are interested in certain aspects of finance, and especially fintech, it makes much more sense from a professional standpoint to remain in Switzerland, the UK, the Netherlands, Singapore or the USA, which are home to vibrant financial services industries and the leading fintech centres (Institute for Financial Services Zug | Lucerne University of Applied Sciences, 2018). Technically, it remains possible for them to work abroad and at some point return home to start a business, but many develop personal networks and start families which, with time, lower the chances of going back (Favell, 2014).

Related to that, the second major problem noted earlier in the chapter is that there are very limited efforts underway to reform education across the region. While at the political level, initiatives have been started and taken onboard, little development has really occurred in terms of making education competitive and relevant for today's generation and the needs of the entrepreneurial ecosystem. Given the time it takes to reshape an educational system and begin seeing results, this may mean that brain drain, which is observable in the region, will remain a severe social and economic issue, including for the future of entrepreneurship. As a former government official from one of the countries notes:

> Education is the most important thing. The cycle of investing in it and seeing something happen is 10–15 years. Even if you do something today you're not going to see an impact, and in the meantime, you're going to keep losing competitiveness.

Deciding on how to reform education is another challenge. While there are some fundamentals that go with overall trends in business and entrepreneurship education, there are specific characteristics and needs relating to context that need to be considered, including the social dynamic, skills and knowledge that are in short supply due to the historical development of the Balkan countries. Here is the view of one of the most successful entrepreneurs in Romania, Florin Talpeş, founder and CEO of the leading provider of cybersecurity and antivirus software Bitdefender:

> We come from countries with a state-owned culture. In all countries in the region, everything was state-owned so before 1990 we did not have entrepreneurship. That, combined with a shortage of what I would call a 'culture of assuming responsibility'—we are not strong on assuming responsibility. The reason for this is that for many generations all the responsibilities lied with the state and the party, our lives were guided. We did not choose an employer, an employer was for life, the career path was clear so there were not many options to take responsibility and choose. The capacity for influencing social rules and practices with the choices one made was very limited. That built a wide culture of not assuming responsibility, as the responsibility for my life was with my boss, the party and the state. Still today, we are struggling to develop this culture of assuming responsibility in children… Education is changing but there are still too many penalties for failure when you are a child…This is where we need to make changes to build a foundation for entrepreneurial culture because entrepreneurship requires assuming responsibility, not only for your life but assuming responsibility for the life of others…. Another thing specific [to the region] is that the number one issue that entrepreneurs here state is that it is difficult for them to find customers. But in reality, this means that they do not

> have sales and marketing skills and knowledge. This is the real issue at hand. This is also because before 1990 everything was state-owned, centralized and planned so we did not need this type of knowledge and skills. There was some need for management capability but not sales and marketing skills… So today, we still have a shortage of business skills for developing an international business… This is another feature of the current entrepreneurial environment that is specific to the region…. In my opinion every ecosystem is unique… In Romania, just like in Bulgaria, the education system is still very STEM-focused, which is producing a lot of STEM skills but still does not do well in entrepreneurial and business skills.

Similar thoughts are shared by lead startup activists in Croatia. Andrej Hanzir from the non-governmental organization Croatian Office for Creativity and Innovation, and lead organizer of the influential startup event LEAP Forum, shared in an interview:

> Because until only 25–26 years ago we were living under socialism and the public sector was really strong, we still don't have this entrepreneurial culture… Related to this, in the whole region the problem is how to accept failure. People are still really afraid of failure…. This is a big problem because if people are afraid of failure, they also do not want to take risks, and entrepreneurship without risk is not possible… We need to have universities that promote entrepreneurship, that offer education in entrepreneurship.

Culture

The region has a few big success stories that carry weight in the global startup arena, most of them concentrated in Croatia, Romania and Bulgaria, but there are many more that demonstrate success in the local and regional context. In Part II of this book, we illustrate these developments with eight selected case studies.

One of the most interesting processes currently taking place in the region is the defining of locally relevant role models. Many entrepreneurs and investors in the region believe that if the region is to become significant on the global startup map, local entrepreneurs need to seek inspiration from global leaders and success stories coming out of the startup cradles of Silicon Valley, London or Tel Aviv. Others suggest that global leaders are inspiring but that it is more important to celebrate the regional success stories of entrepreneurs who have built businesses or found effective solutions to challenges in this specific context and reality.

Stories of entrepreneurial ventures that have initiated and successfully grown businesses around the solving of local or regional problems share several common features: they combine a global mindset and international experience with a strong familiarity with local/regional realities and needs. However, observations made during the extensive research phase accompanying the writing of this book revealed that most people are unaware of regional and even local success stories and they are rarely shared within the startup communities, let alone outside of them.

Entrepreneurs are successful in business because their thinking is different from the rest, according to a successful serial entrepreneur and investor from the region. That said, the socio-cultural setting can be an important facilitator or impediment for entrepreneurship and, as discussed in an earlier section, the overall impression of members of the ecosystem is that the culture in the region is heavily shaped by the historically determined socio-economic dynamics of the last few decades. There are several dominant themes that seem to characterize the specific entrepreneurial culture of the region. First, it is young and is almost entirely a product of the processes in the last 28–29 years since the fall of the Berlin Wall and the dissolving of the Eastern Bloc. This means that many aspects of culture are being formed now in the context of some soul searching between foreign best practices and local specifics. In the words of a prominent regional investor: 'The ecosystem … in the region is still very young and suffers from far too much optimism and far too little knowledge.'

Second, as we have seen, the region formerly operated on a planned economy model, which meant that all economic activity was out of the hands of the individual, which influenced people's attitudes towards entrepreneurship and their abilities to be entrepreneurial. The philosophy that economic decisions were made by the Socialist Party and the government on the basis of what is best for society nurtured discontent towards private initiative. In fact, for generations private initiative was perceived as unethical and associated with freeriding on the back of society. There are still remnants of this thinking within societies in the region today, especially amongst the older generations who lived significant portions of their lives under the old regime and for whom rejecting beliefs that have governed their thinking for so long is nearly impossible.

Third, due to a mix of a culture that has rejected private initiative and promotional behaviour, and the lack of traditions in business and marketing education, there is a significant gap in business thinking and marketing culture. This continues to be a challenge even for countries with more advanced entrepreneurial processes and is certainly a problem for countries with younger startup ecosystems. Michael Bohanes, who spent some time with the startup community in Albania and Macedonia as an entrepreneur-in-residence with support from a Swisscontact programme, observes:

> Here in Eastern Europe people are really bad at presenting their ideas and pitching themselves. They may appear confident on stage, but they are not good at communicating and explaining the essence of their business idea. This applies to all countries in the region.

Jakob Modéer, head of a regional Entrepreneurship programme for the Western Balkans, also notes:

> Many people complain about funding. The problem here is not in the availability of money; it is in the skills. The universal truth is that if you want to develop a business you have to be able to deal with the big areas of ownership, leadership and business organization… You have to be able to see where the money is in the market and then to think about access to capital.

Another significant element of the local culture is associated with the stigma of failure. Unsuccessful efforts are not treated as lessons learned but rather as failures associated with shame. For generations, these attitudes have suppressed the desire to initiate something new or to experiment with bold ideas. Whilst the multitude of startup events, access to behavioural models from leading entrepreneurial ecosystems and discussions around the importance of failures are starting to change that, this view still exists amongst older generations. In the words of Miloš Milisavljević, CEO of one of the successful Serbian startups for smart city technology, Strawberry Energy:

> Fear of failure is traditionally significant here… For the general public in the region, it is still hard to fail because of the overall perception that comes with that and the judgment that people who fail are losers and that they are not good at doing something. This perception might not come from their entrepreneur peers but certainly will come from their parents and society overall. Fear of failure is still high and failure is not embraced.

Andrej Hanzir of the Croatian Office for Creativity and Innovation and LEAP Forum confirms:

> Today we see more cases of celebrating entrepreneurship, so the culture is being formed but it is not there yet. Highlighting good entrepreneurs and good examples, embracing failures as opportunities to learn, accepting risks… we need to speak to and educate young people so they can be even more engaged when it comes to entrepreneurship.

Serbian entrepreneur and investment expert Milos Dosen explains why entrepreneurship is still not perceived as something entirely positive in the region:

> We still have remnants from the socialist past so many people do not have a positive view of entrepreneurship. The problem is that even after the fall of socialism, entrepreneurship was not seen as something positive. The tycoons that gained wealth in unethical ways were referred to as 'entrepreneurs' so in general the population still has a negative attitude towards entrepreneurship.

There is a clear difference between the attitudes of younger and older generations towards entrepreneurship as a career choice. With globalization has come easy access to global role models from entrepreneurial circles and exposure to foreign education, meaning that many young people in the region grow up with a global mindset and values that are similar to those of their peers. During a discussion with several young founders of new startups in Albania, they shared their concerns about the reactions of their parents to their decision to start something of their own:

> I had a 9-to-5 job but I was not happy, and when I told my Mom that I wanted to leave, she told me 'But you are getting paid and you have a paid job! What more could you want?' So, for your parents' generation, it is like this: you have a job, you are safe and that is it. If you want to do something entrepreneurial, that is a bit risky so why do it? I think that the new generation is not like that, it is different.

Another young founder shared her own experience:

> When I told my father I was starting a business, he told me that all the businesses were failing and closing down so why would I think of opening one, and why did I think that my business would be unique and would succeed. It was very difficult to convince him that this made sense. The reason that young people want to start their own businesses is that our generation travels more and we see things abroad that we want to bring to our countries too. We have more information about things too. Our fathers and mothers did not have these experiences. In addition, because many businesses were crushed during the early years of the changes [after the fall of socialism], they got scared and were convinced that they should not even try to start their own business. They simply got scared.

A similar dynamic between generations is still observable in the more experienced ecosystems in the region. Serial Bulgarian entrepreneur Vladimir Rusev shared his own experience and observations about this:

> There is a definite age difference in how entrepreneurship is viewed in society. For example, my parents' mindset has always been that I should get a job in a big company and then the company will train me and pay me a salary, and one day I will retire with this great company. Both my mother and father spent their entire lives in the same career in the same organizations. My path was totally different. I started my first company while I was still at the university and even then my parents kept reminding me that this was very risky and a bit crazy. Later, they saw that I moved out of the house quite quickly and was paying my bills but it was still uncomfortable for them. I see in general that for that older generation, there is a hesitation to

> accept that people can make it on their own. Part of this is because this generation has grown up in times when the state was always very strong and has taken care of them—their education, their job… Entrepreneurship requires the opposite mindset: it means that you are [solely] responsible and you are the one who has to take things into your own hands… With the younger generation, I see a huge difference and I am excited about that. They are willing to take risks, to delay this moment of receiving a higher salary at a big company and taking a couple of years to explore how things might go if they do something they like. They are under less pressure than we were as young people in 1990–1995 to start at big companies and become locked into secure jobs.

Another view on why entrepreneurship is embraced by the younger generation is shared by Larry Biehl, philanthropist and founder of the Bulgarian Entrepreneurship Centre, who thinks that part of the reason we see so much interest in startups and entrepreneurship in the region is that young people are seeing this as an alternative to going abroad for a higher salary:

> A lot of young people have been leaving to get better jobs… Young people are seeking opportunities elsewhere, not in Bulgaria. What we are realizing that a lot of them are wondering how they can stay in the country and be with their families.. and yet make enough money to be able to support this endeavour. This shows they realize they can turn to something more innovative, which is to start their own business and try to scale it up so that they can join the worldwide economy. This is also part of a worldwide phenomenon—more people want to work for themselves rather than large corporations.

The Role of Media

One of the problems that many entrepreneurs and startup specialists in the region talk about is the role of the media and the fact that in some Balkan countries it has failed to celebrate local stories of success or to make successful entrepreneurs into local heroes. Gerti Boshnjaku, an Albanian entrepreneur in the media and publishing domain, commented:

> We are not doing a good job of celebrating entrepreneurs in the media. The role models that Albania knows about are people who maybe got lucky and were simply in the right place at the right moment, or they are people who took advantage of an opportunity and did things that were not right, and that is why they succeeded. This perception of business is not correct and this idea about role model is something that needs to be changed by the media.

In the more advanced startup communities in Croatia and Bulgaria, some media organizations are starting to recognize that they have an important role in establishing successful entrepreneurs as role models. Even though the process has begun, Konstantin Nikolov from Capital newspaper, the leading business publication, which covers startup news, notes:

> I still think the entrepreneurship community is small and many people outside of it think it is a bubble. In Sofia, the ecosystem is pretty well recognized, and there are some examples in the media. I think we need to talk more about the successful examples for the ecosystem to grow.

Ivan Brkin, an entrepreneur himself and one of the most active writers about startup dynamics in Croatia, shares his observations:

> I think entrepreneurs have a place in the media and are well known. I think the media covers startups very well although in most cases the media talks naïvely about startups. I would say that in Croatia, the media and institutions have understood that they need to take a role and they have to give their support. Now, tech startups are the champions of Croatian business and that is the result of a lot of people working very hard, both from institutions and the media, trying to support this.

A review of the historical media coverage across the different countries in the region reveals that there is indeed an increase in the amount of attention that the media devotes to startup news and entrepreneurship topics. In the more advanced ecosystems of Bulgaria, Croatia, Romania and Slovenia, there are TV programmes as well as regular sections in business newspapers and magazines devoted to the topic. In addition, numerous blogs and social media platforms serve as key media channels for the regional community.

References

Audretsch, D. B., Baljeâ, S., Carree, M., Klomp, L., Snijders, J., Van Stel, A., … Verhoeven, W. (2000). Capitalism and democracy in the 21st century: From the managed to the entrepreneurial economy. *Journal of Evolutionary Economics, 10*(1–2), 17–34. https://doi.org/10.1007/s001910050003

CIA Factbook. (n.d.). *The World Factbook*. Central Intelligence Agency. Retrieved May 29, 2018, from https://www.cia.gov/library/publications/the-world-factbook/geos/si.html

EIF. (2015). *JEREMIE: A new way of using EU structural funds to promote SME access to finance via holding funds*. Luxembourg. Retrieved from http://www.eif.europa.eu/news_centre/publications/Jeremie_leaflet_files/jeremie_leaflet_en.pdf

Fastmetrics: Internet Speeds by Country—Fastest Internet in the World Map. (n.d.). Retrieved June 2, 2018, from https://www.fastmetrics.com/internet-connection-speed-by-country.php

Favell, A. (2014). The new face of east–west migration in Europe. *Journal of Ethnic and Migration Studies, 34*(5), 701–716. https://doi.org/10.1080/13691830802105947

Hallet, M. (1997). *National and regional development in Central and Eastern Europe: Implications for EU structural assistance*. Retrieved from http://ec.europa.eu/economy_finance/publications/pages/publication11247_en.pdf

Hatfield, I. (2015). *Self-employment in Europe*. London. Retrieved from https://www.ippr.org/files/publications/pdf/self-employment-Europe_Jan2015.pdf

Institute for Financial Services Zug | Lucerne University of Applied Sciences. (2018). *Global FinTech Rankings | Thomson Reuters*. Retrieved from https://innovation.thomsonreuters.com/en/labs/portfolio/global-fintech-rankings.html#/

Kluczynski, J. (1980). *Trends of educational policy in socialist countries. Supportive study for the preparation of document ED-80/MINIDEUROPE/3 for the conference of ministers of education of members states of the Europe region due to be held in Sofia (Bulgaria) in June 1980* (No. UNESCO ED/EPP/S.76 Ed-80/WS/51). Retrieved from http://unesdoc.unesco.org/images/0003/000389/038902Eb.pdf

Lokshin, M., & Fong, M. (2000). *Child care and women's labor force participation in Romania* (Policy Research Working Papers). Washington, DC: The World Bank. https://doi.org/10.1596/1813-9450-2400

Meurs, M., & Giddings, L. (2004). *The demise of state run child care in Bulgaria: Causes and implications* (Department of Economics Working Paper Series No. 200406-01). Washington, DC. Retrieved from http://www.american.edu/cas/econ/workpap.htm

Move.bg. (2017). *2017 EDIT national report on digital micro, small and startup business in Bulgaria*. Sofia. Retrieved from http://edit.bg/wp-content/uploads/2017/11/innovationship-2017-14-11-EN-print.pdf

Mylonas, P., & Tzakou-Lambropoulou, N. (2016). *Survey of Greek SMEs second half of 2015 special issue: Capital controls*. Athens. Retrieved from https://www.nbg.gr/english/the-group/press-office/e-spot/reports/Documents/SMEs_Survey_2015H2_EN.pdf

Peshev, N. (2017, October 24). A family company received funding straight from Elon Musk. *Medium*.

Silcox, S. C., Lobron, K., & Nathanson, N. (2005). *Assessment of the Bulgarian enterprise growth and investment project (Final report)*. Washington, DC. Retrieved from http://www.oecd.org/countries/bulgaria/35893427.pdf

4

Balkan Startups and the Entrepreneurs Behind Them

Introduction

Early-stage ventures and their entrepreneurial founders are a fascinating topic for research, but more often than not researchers approach the study of entrepreneurs by applying a number of general theories and frameworks that help them to simplify the enormous richness of detail that surround the stories of each and every entrepreneur. And there certainly is an abundance of fascinating stories of entrepreneurship in the Balkans.

One of the most inspiring entrepreneurial success stories from the Balkans comes from Croatia, where, in 2011, Mate Rimac developed the world's first electric supercar, the Concept One: a car with 1088 horsepower that can reach 100 km/h in 2.8 seconds and that is only comparable to other supercar brands such as Pagani and Bugatti (*Compass Magazine*, 2018). The car has 24 patented innovations and is made from parts built entirely in Croatia (Rimac Automobili, 2018).

Rimac's entrepreneurship story began as a hobby in 2007, when he burned out the engine of his BMW and decided to replace it with his own electric engine. His dream was to create the first electric sports car,

V. Andonova et al., *Entrepreneurial Ecosystems in Unexpected Places*,
https://doi.org/10.1007/978-3-319-98219-9_4

and by 2009 he had founded his company, Rimac Automobili. Starting the company in Croatia was a huge challenge for Rimac, since he received no financial support from the government or local investors, and there was no local market for his product (CTV News, 2016). Furthermore, the motor industry in Croatia was very small, with no automotive clusters, so there were very few people who had automotive experience who could help him. Nevertheless, the company has grown since its creation in 2009, despite the difficulties it faced during its early stages. Rimac chose to develop the company in Croatia rather than moving elsewhere, and by 2016 the company had over 300 employees with two factories in Sveta Nedelja.

Since the creation of the Concept One in 2011, the company has evolved, and it has built a global leadership position in electric vehicle technology (Bach & Zubrinić, D., 2013). Its core business has become the sale of high-end electric driving system components to other car companies such as Aston Martin, Jaguar and Renault. The company's technology is also being used in other products, like wheelchairs and bikes, which shows how Rimac has pushed the boundaries of his original dream of creating an electric supercar to innovate and serve new markets. Rimac Automobili's supercars have become a showcase for what the company can offer in terms of technological innovation, and the company's business model has helped it become a successful entrepreneurship story.

Gjirafa presents another impactful story but in a different domain.

Albanian is one of the most difficult and unique languages spoken in the Balkan region. It belongs to the Indo-European language family, with influences from Romanian, Greek and even Turkish (Thompson, 2016). It is spoken by 12 million people, not only from Albania and Kosovo but in several other countries in the region. The complexity and uniqueness of the language has led the Albanian internet community to face difficulties when using regular search tools, such as Google, to browse the web efficiently. Mergim Cahani, founder and CEO of Gjirafa, saw this as an opportunity to build a comprehensive web platform for the Albanian-speaking community (Reback, 2016).

Gjirafa was founded in 2013 in Kosovo, where it began as a full-text web search engine focused entirely on the Albanian market and specializing in the Albanian language. One of Gjirafa's main goals was to make information accessible to Albanian speakers, information which is not easily accessible or simply did not exist yet on the internet. It has since grown into a full-scale web platform, with e-commerce solutions, a virtual marketplace, job listings, advertising, news and a lot of content that is uniting the Albanian-speaking community. This all took place as the company founders developed a deeper understanding of the regional market and improved content availability. For Cahani, its founder, Gjirafa would become the home page of the Albanian-language web, providing relevant services and easy access to information for the community (Waldo, 2014).

Gjirafa was helped by two accelerators during its growth stage. First, it received support from Startup Yard, a Czech accelerator which provided mentorship and contacts in the region that supported its development. Second, in 2016, Gjirafa participated in a Series A funding round and received US$2 million from Rockaway Ventures, a Silicon Valley/Czech Republic venture fund (O'Hear, 2016). This funding helped grow Gjirafa by digitizing and indexing information from the region, and further expanding its services (Crunchbase, 2018).

Gjirafa is a good example of a company from the Balkans that has been able to innovate and fill a gap in the regional market, rather than aiming at the global market, as Rimac has done. The company has seen a lot of growth since its inception, and coming from the smallest country in the region, Kosovo, which also retains some of the remaining geopolitical challenges in the region, it illustrates the strength of the local entrepreneurial spirit.

The stories differ, but entrepreneurs are the same all over the world. These are words we often hear when speaking with mentors visiting the Balkans after they have delivered their mentoring sessions to local startups. We believe this is not necessarily the case: our view is that every region, every country, and even every city, has its own peculiar environment and history that, to a large extent, determine and shape the motivation of founders, their ambitions and their approach to entrepreneurship. In this chapter, we perform a quantitative analysis of the data available from the Entrepreneurship Database Program at Emory University and on the primary data of application and portfolio samples of both Eleven and LAUNCHub's accelerator venture funds. An interpretation of the results reported here relies on assumptions and conventions that have been established in different latitudes, so a critical and reflective analysis of the profile of a Balkan startup and its founders is possible only if the quantitative data is complemented with a series of structured interviews with startup founders from the region, and these form part of the second half of this book.

Balkan Entrepreneurial Ecosystems: A First Approximation

The startup ecosystems in the countries we cover here are relatively young, most of them starting and growing rapidly around 2012, when startup funding became available with the launch of Eleven and LAUNCHub. This does not mean to say that there was no such activity before 2012, but the rapid growth of the startup movement was catalysed and fuelled to a great extent by this event, so it is from that date that a more defined profile of Balkan startups and entrepreneurs started to form. A more detailed historical account, as well as opinions and testimonials of key participants, is provided in Chap. 3. This section is only a first glance at a process that has been silently transforming Balkan countries.

In the late 1990s and early 2000s, the region was regarded as a sweatshop, providing low-cost engineering and development services to US and Western Europe-based clients. It gradually evolved into a business process outsourcing destination, with many international corporations setting up their own development arms there, which in turn gave rise to the accumulation of expertise and a better understanding of how technological service companies operate. Shortly thereafter, local entrepreneurs began to set up their own service businesses in the technology sector. Angel investor networks were established in some Balkan countries, but the number of investments made by these ranges from zero to fewer than ten, depending on the country. Regular meetups and startup events were sporadic, but slowly gained momentum.

The landscape changed drastically in 2012, when the two early-stage investment funds started their operations out of Sofia: Eleven, with €12 million under management, and LAUNCHub with €9 million. The funds' strategy was to invest between €25,000 and €200,000 in early-stage companies, incorporated in Bulgaria, with the focus on founders originating from the region. Despite the fact that the funds were entirely return-minded (meaning they had to provide a return to their investors), they also had a developmental intent, as one of the Key Performance Indicators (KPIs) of the funds was the number of invested startups. Furthermore, the investment horizon of both funds was December 2015, so they only had a three-year investment window.

Given the fact that some 20 years earlier private entrepreneurship had been forbidden, and despite the fact that the 20–30-year-olds from the region had spent the greater part of their lives in a free market economy, there was virtually no tradition in entrepreneurship, let alone high-growth technological entrepreneurship. Therefore, the two funds had the hard task of educating the market, inspiring local entrepreneurs to quit their jobs and start companies, sourcing and selecting, investing in, and supporting hundreds of companies in a place where there was little to no previous tradition of an entrepreneurial culture. The first thing the funds did was to make a lot of noise on traditional and social media about the available opportunity, and they got in touch with anyone in the region who was involved one way or another with entrepreneurship—be it event or conference organizers, bloggers or influencers. This increased awareness created a snowball effect, with more startup events and competitions appearing all over the region, which provided opportunities for investors and evangelists to preach and lecture about entrepreneurship. In one short year, thousands of people were inspired, or at least influenced, by this movement, and this process still continues today.

The availability of funding and the growing number of funded startups opened the doors for a number of co-working spaces, conferences, pitch competitions and startup events. The speed at which the word was spread and people were inspired was astonishing. In the very first application window, when the funds opened in June 2012, close to 1000 applications were submitted. About 6500 applications for investment were submitted to both accelerators in the three years they were actively investing, adding up to a total of about 20,000 people (at an average of three persons per team) considering starting their own business. Before we look more closely at this data, we will take a comparative approach and inspect a series of founder and startup characteristics across different geographical regions, with a particular focus on the Balkans.

Balkan Ventures in the Entrepreneurship Database Program at Emory University

The Entrepreneurship Database Program collects information from entrepreneurs who apply to accelerator programmes around the world. The methodology behind the data relies on the re-surveying of entrepreneurs selected by the accelerator programmes, as well as those who were not selected, on a six-monthly basis. The primary objective of this data collection effort is to support accelerator programme managers, report to funders and present evidence about the programme's impacts to all remaining stakeholders. We use the data collected up to 2017 to characterize startups from the following Balkan countries on which data was available: Bulgaria, Romania, Serbia, Greece and Slovenia. Table 4.1 contains a list of countries where Balkan, Baltic and Central European startups were reported to be headquartered. Ten per cent of the Balkan ventures were headquartered in the USA or UK, but most of them chose their home region in which to locate their headquarters.

The Entrepreneurship Database Program had as few as 24 Balkan ventures founded between 2009 and 2016. Most of them, but one, were for-profit organizations (see Table 4.2).

Of these 24 ventures, none was selected to participate in an accelerator programme. As with the Baltic and Central European ventures, the main operating model of the early-stage Balkan companies was *Services* (see Table 4.3).

Balkan ventures do not have a dominant or a narrow sector focus, even though several areas of specialization can be identified, including *Information and Communication Technology* (ICT) and *Tourism* and *Environment* (see Table 4.4). To aid comparison, we show in Table 4.5 the industry focus of a sample of applicants to LAUNCHub, the accelerator venture fund. Using this alternative source of data, a broad dispersion across industries is confirmed, with a few areas of specialization including social media, marketing analytics, e-commerce and sports and travel.

Table 4.1 Stated headquarters and the origin of the founders

Headquarters in	Balkan states (%)	Baltic states (%)	Central Europe (%)
Albania	4	0	0
Bulgaria	10	0	0
Chile	2	8	25
Estonia	2	46	0
Greece	40	0	0
Hungary	0	0	30
Latvia	0	8	0
Lithuania	0	23	0
Poland	0	0	30
Romania	10	0	0
Serbia	4	0	0
Slovakia	0	0	15
Slovenia	13	0	0
Switzerland	2	0	0
FYR Macedonia	2	0	0
Ukraine	0	8	0
UK	4	0	0
USA	6	8	0

Balkan states represented are Albania, Bulgaria, Romania, Serbia, Greece, Croatia FYR Macedonia and Slovenia

Baltic states represented are Estonia, Latvia and Lithuania

Central European states represented are Hungary, Poland and Slovakia

Data provided by the Entrepreneurship Database Program at Emory University, supported by the Global Accelerator Learning Initiative. Data collected between 2013 and 2017

Table 4.2 Legal status of startup ventures

Legal status	Balkan states	Baltic states	Central Europe
For-profit	83.3	92	90
Undecided	6.30	8	10
Non-profit	8.30	–	–
Others	2.1	–	–

The Balkans are represented by Albania, Bulgaria, Romania, Serbia, Greece, Croatia FYR Macedonia and Slovenia

Baltic states are Estonia, Latvia and Lithuania

Central Europe represented by Hungary, Poland and Slovakia

Data provided by the Entrepreneurship Database Program at Emory University, supported by the Global Accelerator Learning Initiative. Data collected between 2013 and 2017

Table 4.3 What is the current operating model of your venture?

Model	Balkan states (%)	Baltic states (%)	Central Europe (%)
Services (S.)	54	62	60
Financial Services	13	8	5
Manufacturing (M.)	8	15	5
Combination (M. & S.)	8	7.7	0
Combination (M. & Distribution)	6	0	5
Combination (S. & Retail)	2	0	5
Combination (S. & Distribution)	0	8	0
Combination (M., Packing, Distribution & Retail)	2.1	0	5
Unsure	4.2	0	5
No answer	2.1	0	10

Balkan states represented are Albania, Bulgaria, Romania, Serbia, Greece, Croatia, Macedonia and Slovenia

Baltic states represented are Estonia, Latvia and Lithuania

Central European states represented are Hungary, Poland and Slovakia

Data provided by the Entrepreneurship Database Program at Emory University, supported by the Global Accelerator Learning Initiative. Includes author's calculations

Data collected between 2013 and 2017

Entrepreneurial ventures list *community development* and *income and productivity growth* as their top priorities when it comes to producing an impact (see Table 4.6). The fact that community development still features highly in their agenda is notable, and we will explore how Balkan ventures combine for-profit and societal purpose in their business models in more detail in Chap. 5.

Most Balkan ventures in the 2017 dataset reported specific profit target margins of *above* 20%, with 10% reporting profit target margins of 16–20%, 20% reporting a profit target of 15% or below, and 30% did not have a specific profit margin target. The overall pattern is similar to that presented by Baltic startups and markedly different from the pattern presented by Central European ventures, which tend to be less ambitious and report lower profit target margins (see Table 4.7).

However, when it comes to financial goals, the data up to and including 2016 shows a somewhat different picture. Only 11% of Balkan ventures expect to cover their costs and make a profit. Baltic ventures appear much

Table 4.4 Which primary sector is being impacted by your venture's activities?

Primary sector	Balkan states (%)	Baltic states (%)	Central Europe (%)
Agriculture	4	8	10
Culture	2	0	5
Education	6	31	0
Energy	6	0	5
Environment	8	0	0
Financial Services	2	8	10
Health	6	31	10
Housing Development	2	0	0
Information and Communication Technology	13	23	20
Supply Chain Services	4	0	0
Technical Assistance Services	0	0	5
Tourism	10	0	15
Others/not specified	35	0	20

Balkan states represented are Albania, Bulgaria, Romania, Serbia, Greece, Croatia FYR Macedonia and Slovenia
Baltic states represented are Estonia, Latvia and Lithuania
Central European states represented are Hungary, Poland and Slovakia
Data provided by the Entrepreneurship Database Program at Emory University, supported by the Global Accelerator Learning Initiative. Includes author's calculations
Data collected between 2013 and 2017

more ambitious (or optimistic), as the percentage of those who expect to cover costs and make a profit is twice as high (see Table 4.8).

Additionally, Balkan ventures appear to be extremely optimistic about their average expected investment by source. In respect of every source of investment, Balkan startups have at least double the expectations of Baltic and Central European ventures (see Table 4.9).

When we turn to the very first startups for which data was internationally collected, we find that, up until 2016, only 17 Balkan companies (out of a total of 24 by this cut-off point) considered themselves invention-based, according to the information in the Entrepreneurship Database Program. Ten of them have at least one patent, copyright or trademark. Only nine ventures reported any prior year revenues, with an average of US$14,020.83. Four of them accomplished their profit margin target of more than 20%, while seven ventures reported losses in the previous year;

Table 4.5 Industries reported by LAUNCHub applicants

Industry	Frequency	Per cent (%)
Clean tech	7	1
Dev. tools	5	1
Education	30	6
Entertainment	13	2
Fintech	26	5
Food and Health	26	5
Gaming	18	3
HR and legal	16	3
Hardware/drones	16	3
Marketing analytics	45	8
Med tech	28	5
Others	113	21
Saas/tools	40	7
Sales/CRM	11	2
Social media	66	12
Sports/travel	48	9
e-Commerce	34	6

Source: Authors' calculation from LAUNCHub application data
Data collected between 2014 and 2015

four had a small but positive margin of 0–5%. Twelve ventures had at least one full-time employee in the previous year.

Owner investments since founding their firm (not counting the six ventures whose owner investment was null) totalled on average US$36,384.36. Six ventures had received equity financing, from angel investors and other sources, reporting an average of US$78,112.50. Three other companies had used debt from family members with an average of US$21,900 per startup. Only one company received founding from a philanthropic source, equal to US$12,000. To secure the continuity of their operations, 14 ventures planned additional equity investment of an average of US$403,278 for the subsequent 12 months. Four of them planned to take on debt in the next 12 months and 7 for the next 3 years with averages of US$151,975 and US$458,571, respectively. Seven ventures expected to receive investment from philanthropic sources in the next 12 months, with an average of US$178,285.

Nineteen of the ventures had at least two founders, who were, on average, 33 years old. Of the 53 founders, only 7 were female.

Table 4.6 Which of the following impact objectives does your venture currently seek to address?

Impact area	Balkan states (%)	Baltic states (%)	Central Europe (%)
Community development	13.6	10.5	19.4
Access to information	6.8	15.8	13.9
Health improvement	6.8	10.5	8.3
Access to education	4.9	15.8	5.6
Disease-specific prevention and mitigation	1.0	5.3	2.8
Affordable housing	1.9	–	–
Biodiversity conservation	1.0	–	–
Access to financial services	–	–	2.8
Capacity building	1.00	–	2.8
Agriculture productivity	2.90	–	8.3
Human rights protection or expansion	–	5.3	–
Support for women and girls	–	–	2.8
Employment generation	7.8	–	5.6
Energy and fuel efficiency	3.9	–	5.6
Sustainable land use	2.90	–	2.8
Conflict resolution	1.0	5.3	–
Food security	1.90	–	2.8
Equality and empowerment	3.9	–	2.8
Generate fund for charitable giving	1.0	–	–
Income/productivity growth	14.6	21.1	2.8
Sustainable energy	2.9	–	2.8
Pollution prevention and waste management	7.8	–	–
Support for high-impact entrepreneurs	1.0	–	–
Water resources management	1.9	–	–
Others	8.7	10.5	8.3

Balkan states represented are Albania, Bulgaria, Romania, Serbia, Greece, Croatia FYR Macedonia and Slovenia

Baltic states represented are Estonia, Latvia and Lithuania

Central European states represented are Hungary, Poland and Slovakia

Data provided by the Entrepreneurship Database Program at Emory University, supported by the Global Accelerator Learning Initiative. Data collected between 2013 and 2017

Table 4.7 Anticipated profit margin for firms

Target margin	Balkan states (%)	Baltic states (%)	Central Europe (%)
More than 20%	43	40	27
16%–20%	10	0	20
11%–15%	7	10	0
6%–10%	10	0	7
0%–5%	2	0	0
None	29	50	47

Balkan states represented are Albania, Bulgaria, Romania, Serbia, Greece, Croatia, Macedonia and Slovenia
Baltic states represented are Estonia, Latvia and Lithuania
Central European states represented are Hungary, Poland and Slovakia
Data provided by the Entrepreneurship Database Program at Emory University, supported by the Global Accelerator Learning Initiative. Data collected between 2013 and 2017

Table 4.8 Anticipated financial goals for firms

Financial goal	Balkan states (%)	Baltic states (%)	Central Europe (%)
Cover costs and earn some profit	11	23	11
Cover costs	89	77	89

Balkan states represented are Albania, Bulgaria, Romania, Serbia, Greece, Croatia FYR Macedonia and Slovenia
Baltic states represented are Estonia, Latvia and Lithuania
Central European states represented are Hungary, Poland and Slovakia
Data provided by the Entrepreneurship Database Program at Emory University, supported by the Global Accelerator Learning Initiative. Data collected between 2013 and 2017

Baltic Ventures

This sample consists of seven ventures founded between 2010 and 2016 from the following countries: Estonia, Latvia and Lithuania. Six of them considered themselves to be for-profit organizations, and the remaining one was undecided. Just one of them had participated in an accelerator programme. Two of the ventures indicated *Health* as their primary sector, another two had *Information and Communication Technology (ICT)* as their primary sector, and the others had *Agriculture*, *Financial Services* and *Education* as their primary sectors. Three of the ventures reported

Table 4.9 Average expected investment of firms by source in US dollars

	Balkan states		Baltic states		Central Europe	
Source	Next 12 months	Next 3 years	Next 12 months	Next 3 years	Next 12 months	Next 3 years
Equity	$168,555.98	$3,863,832.96	$77,692.31	$400,769.23	$71,181.30	$270,000.00
Debt	$26,526.85	$103,175.63	$9,615.38	$27,692.31	–	–
Philanthropic	$35,143.67	$35,354.06	$5,000.00	$14,000.00	$26,500.00	–

Balkan states represented are Albania, Bulgaria, Romania, Serbia, Greece, Croatia FYR Macedonia and Slovenia
Baltic states represented are Estonia, Latvia and Lithuania
Central European states represented are Hungary, Poland and Slovakia
Data provided by the Entrepreneurship Database Program at Emory University, supported by the Global Accelerator Learning Initiative. Data collected between 2013 and 2017

Services as their operating model; two of them reported *Manufacturing*; one reported *Financial Services*, and one reported a combination of *Distribution* and *Services* as their operating model. Only one of the ventures reported having a specific profit margin target of 20% or more; another had a target of 11–15% and the rest didn't have any specific target. Five of the ventures had non-financial motives, such as creating social and/or environmental impacts. Only four of the ventures had at least one patent, trademark or copyright, while five of them still considered themselves as *invention-based* ventures.

Two ventures reported having revenues in 2013, with an average of US$23,500. Three of the ventures reported having small but positive profit margins of 0–5% in 2013. Two ventures reported having owner investments, with an average of US$310,000. Two of the ventures reported having equity financing since they were founded, but only one reported the amount (US$40,000). Only one venture reported having debt (US$25,000). Two of the ventures had received funding from philanthropic sources (US$22,500 on average). In order to secure growth, 2 of the ventures planned to take on debt in the following 12 months of US$37,500 on average and an average of US$75,000 in the following 3 years. Three ventures expected additional equity financing in the next 12 months (US$270,000 on average).

The venture founders had an average age of 31 years and only one of the founders was female. Four of the ventures had three founders, one of them had six founders, another had two founders and the one had just one founder.

Central European Ventures

This sample consisted of 15 ventures founded between 2015 and 2016 in Hungary, Poland and Slovakia. Fourteen of them explicitly identified themselves as for-profit operations. Only four of them had participated in accelerator programmes. The primary industrial sectors for the ventures were *Information and Communication Technology* for three ventures; *Financial Services* for two of them; *Health* for two of them; *Tourism* for two of them; and *Agriculture, Culture* and *Technical Assistance Services* for the remaining ventures. Five ventures used *Services* as their operating model, one of them provided *Financial Services*, another had *Manufacturing* as their operating model and two of them described their operating model as a combination of several models (*Manufacturing, Retail, Packing, Distribution* and *Services*). Three of the ventures reported having a specific profit margin target of 20% or more; two of them had a target of 16–20%, another one had a target of 6–10% and the remaining eight didn't report any specific profit margin target. Ten of the ventures had non-financial motives such as creating social and/or environmental impacts. Only one of the ventures had at least one patent, trademark or copyright, but eight of them considered themselves to be invention-based businesses.

Only one venture reported having revenues of US$1500 since they were founded. One of the ventures reported losses, and three of them had small but positive profit margins of 0–5% in 2013. Owner investment was on average US$5550 for the four ventures that reported information on this item. Four ventures reported having financing from angels, venture capitalists or other sources, with an average of US$215,750 invested. None of the ventures had taken on debt or received funding from philanthropic sources. In order to secure growth, three of the ventures expected additional equity financing in the next 12 months of an average of US$53,333, and in the next three years of an average of US$966,666. Another venture expected funding from a philanthropic source in the next 12 months to the tune of US$20,000.

The venture founders were 30.5 years old on average and 7 of the 28 founders were female. Six of the ventures had two founders, four of them had three founders, and the remaining five ventures had one founder each. Just one of the accepted ventures actually finished the acceleration programme.

This comparison between the first startup ventures and founders in the Balkans, Baltic states and Central European countries present in the Entrepreneurship Database Program illustrates that there are few areas that point to significant regional specificities among the startups in these regions. Certainly, one explanation of this is the fact that these are small-number comparisons (24 Balkan, 7 Baltic and 15 Central European ventures). Still, it is useful to assess the characteristics of these entrepreneurial initiatives in comparison with the entrepreneurial ventures in the full dataset which, in 2016, consisted of 5224 entrepreneurs who applied to 82 accelerators and support programmes.

Most of the ventures (4068) in the full global dataset were for-profit and were, on average, younger (2.5 vs. 3.8 years) than the non-profit group (600). The least common way of initial financing was taking on debt. Only 14.5% reported taking on debt to finance their startup (the average of these types of investments was US$32,250); 17.9% reported receiving equity investment (with an average of US$56,000), while a larger number received philanthropic contributions, representing 27% of the total (investment average of US$20,000). Balkan ventures, however, reported very small philanthropic contributions and greater reliance on both equity and debt. This may be for social and cultural reasons, as philanthropic activity is new for most post-Communist societies.

Europe and Central Asia did relatively well in their ability to access equity financing (21.6%), being overtaken only by North America (23%) and Middle East and North Africa (27.6%). Despite the fact that the number of European ventures in the dataset is very small (only 116 ventures), European ventures and Balkan ventures in particular do not appear to be at a significant disadvantage in terms of access to equity financing in comparison to other geographical regions.

Equity investment was more common in ventures oriented to Financial Services (29.3%) and Energy (22.8%), and less common in Housing Development (10.6%) and Technical Assistance Services (11.6%). Ventures dedicated to the Artisanal sector (71%) and the Supply Chain sector (59.6 %) performed better in reporting any prior year revenues.

A large share of the total sample (38%) reported ambitious profit margin aspirations (margins of >20%). Almost half of the ventures had no women as founders. Teams consisting of men only had more likelihood of accessing equity investments (21% m. vs. 14.5% f.). However, teams with women as founders did significantly better in the reported prior year revenues (51.4% f. vs. 41.7% m.). Additional analysis of the data shows that this bias against women persists in both inexperienced and experienced teams and characterized the Balkan ventures in the dataset too.

Fifty-five per cent of the ventures in the sample had at least one team member with some prior entrepreneurial experience. These teams were provided with better opportunities to access equity investments (20.9% vs. 14.1%) and performed better in the reported prior year revenues (47.5% vs. 45.2%).

Ventures that own intellectual property, such as patents, copyrights or trademarks, have better access to equity investments (26.2% vs. 11.6%) and report higher revenues (52% vs. 42.4%).

Thirty per cent of the sample participated in an accelerator programme. These teams reported fewer equity investments (14.6% vs. 18.9%) but performed significantly better in the reported prior year revenues (53.2% vs. 44.4%). In this regard, Balkan (as well as Baltic and Central European) ventures appear to be at a significant disadvantage. However, more recent and dynamic entrepreneurial developments have rapidly changed the exposure of entrepreneurs in the region to acceleration programmes and all their associated benefits, as testified by founders in the detailed accounts of their entrepreneurial experiences in the second part of this book.

In fact, there are multiple ways in which to measure the ongoing progress of a startup, and it is not obvious to founders nor to investors whether there is one common methodology for accomplishing this task. The next section looks at an ambitious effort to build a framework to measure the progress of new ventures: the Startup Genome Report.

Startup Genome Report

The Startup Genome Report was launched in 2012 by Max Marmer, Bjoern Herrmann, Ertan Dogrultan and Ron Berman, who represent the Startup Genome founding team, in collaboration with Berkeley and Stanford affiliated faculty.

The Startup Genome Report was built around a basic startup lifecycle model whose objective was to present a framework to effectively assess startups. The model was validated with convenience samples collected from 650+ US-based internet startups concentrated in the Silicon Valley area. The model analyses the venture progress during six stages of development: Discovery, Validation, Efficiency, Scale, Profit Maximization and Renewal or Decline.

The first stage, Discovery, was found to last an average of five to seven months. During this stage, the founding team is formed and the startup's value proposition is established by exploring whether the venture actually solves a problem and whether people are interested in the proposed solution. The next stage, Validation, was found to last an average of three to five months. During this stage, startups go through an early validation of their product or service by potential users. Normally, progress metrics are introduced during this stage and a refinement of the core features is carried out. Necessary pivots from the founders' initial strategy are undertaken. Efficiency is the next stage and lasts an average of five to six months. During this stage, the business model and the value proposition are refined in order to improve the customer acquisition process. The Scale stage follows. This stage lasts an average of seven to nine months, and this is the time when startups seek to grow aggressively, with improvements in all their processes, all geared to massive customer acquisition. The final two stages, Profit Maximization and Renewal (or Decline), follow, but these stages are more relevant to mature organizations and are not applicable when the focus is early-stage ventures.

The essential premises on which the lifecycle model of startup development relies are as follows: (i) learning is a fundamental unit of progress for startup; (ii) startups develop via discrete stages of development and each stage can be measured; (iii) there are different types of startups that progress through the lifecycle stages differently.

Using this lifecycle model, an analysis of traditional performance indicators revealed that moving through the initial stages took more time than many first-time founders had expected. The founders' natural impulse to move rapidly to subsequent stages of the lifecycle resulted in inconsistent learning processes that resulted in undesirable consequences for the early-stage ventures. A key insight was that investments grew with each subsequent stage (except between the Validation and Efficiency stages, where investors focused on efficiency), and the biggest rise occurred during the Scale stage. Team size grew the most during Scale-up, while it stayed relatively constant prior to this stage.

According to the survey results that were based on self-reported data by startup founders, the main factors that led to competitive advantages at each different stage were the following:

- *Discovery stage*: Intellectual property (IP) and technology
- *Validation stage*: partners and insider/privileged information
- *Efficiency stage*: traction, IP and insider/privileged information
- *Scale stage*: IP, technology and traction

Startup founders also reported on the main challenges they faced. During the Discovery stage, founders' main concerns were customer acquisition and overcapacity. During the Validation stage, they reported that customer acquisition, product–market fit and problem solving were the main challenges. The main challenges during the Efficiency stage were customer acquisition, team building and fundraising, while during the Scale stage, founders identified customer acquisition and team building as key challenges. It is obvious that customer acquisition is by far the biggest challenge throughout the four stages and that it is a process that requires a lot of time and development from the founding team.

Consistency, in other words, startups that learn incrementally and thus move sequentially from one to the next developmental stage, was considered a key element for success. Startups that didn't move consistently (sequentially) through the stages performed worse. Many of these tried to leapfrog to the next stages without effectively completing the learning process associated with the earlier ones. Premature scaling was a very common source of inconsistency; this happens when a startup attempts to scale without validating whether people actually want the product. Data showed that inconsistent startups experienced a different process regarding pivots, important changes in the way in which value was added. Inconsistent startups either did not pivot at all or pivoted very frequently. Also, inconsistent startups attracted less funding than consistent ones. Moreover, consistent startups' employee base grew five times faster than that of inconsistent ones.

Essentially, the main message was that the process of developing successful startups differed because of their different learning abilities, attributable in part to the flexibility and adaptability shown by the founding team, which was seen as an important determinant of success. Unsurprisingly, the report concluded that startups were much more successful if they had helpful mentors, since Scale-stage startups were unable to raise funds without helpful mentors. Also, the monitoring of key metrics was considered to increase the speed of learning, because both founders and investors knew for sure what was working and what was not, facilitating the decision-making process, which led to better results. The report revealed that startups that monitored metrics not only had better user growth but were also 61% more likely to attract investment. Nonetheless, it was difficult to come up with and craft metrics that were actually useful for the assessment of early-stage startups. There are currently no predetermined universally applicable measurements; nor is there an established template of successful startups standing to serve as a basis for new entrepreneurship efforts. What is certain is that measuring certain aspects of a startup results in more successful development.

The Startup Genome of Balkan Ventures

How would Balkan ventures compare to the ideal of a continuously learning startup as described in the Startup Genome report? We tried to answer this question by applying the Startup Genome instrument to the startups in the Eleven portfolio and, in what remains of this chapter, we present the profile of an early-stage venture that was successful in securing seed-stage funding between 2012 and 2015 in the Balkans. We applied an extensive survey instrument that was answered by 56 startups from the Eleven investment portfolio.

Using application data from between November 2012 and December 2015, Table 4.10 describes the country of origin of the applicant teams. While more than 60% of the applications came from Bulgaria, it is clear that the financing opportunity was recognized regionally. In fact, this regional attractiveness increases for later batches of startup applicants, revealing a consolidation of regional reputation but also representing, some argue, a symptom of a depletion of the local pool of fundable startup teams.

Table 4.10 Stated headquarters of Eleven application sample

Headquarters	Per cent (%)	Frequency
Bulgaria	60.19	1332
Serbia	7.28	161
Romania	3.52	78
Croatia	2.71	60
UK	2.26	50
USA	2.26	50
Macedonia	2.17	48
India	1.72	38
Greece	1.63	36
Italy	1.45	32
Slovenia	1.31	29
Russia	1.22	27
Germany	0.86	19
Armenia	0.68	15
Ukraine	0.68	15
Hungary	0.63	14
Netherlands	0.59	13
Spain	0.59	13
Montenegro	0.54	12
France	0.50	11
Turkey	0.50	11
Argentina	0.45	10
Portugal	0.45	10
Israel	0.41	9
Albania	0.36	8
Georgia	0.36	8
Estonia	0.32	7
Austria	0.27	6
Belgium	0.27	6
Brazil	0.27	6
Czech Republic	0.23	5
Japan	0.23	5
Latvia	0.23	5
Canada	0.18	4
Chile	0.18	4
Denmark	0.18	4
Finland	0.18	4
Malaysia	0.18	4
Mexico	0.18	4
Moldova	0.18	4
Poland	0.18	4
Australia	0.14	3

(*continued*)

Table 4.10 (continued)

Headquarters	Per cent (%)	Frequency
Cyprus	0.14	3
Kazakhstan	0.14	3
Lithuania	0.14	3
Colombia	0.09	2
Jordan	0.09	2
Kyrgyzstan	0.09	2
Nigeria	0.09	2
Slovakia	0.09	2
Sri Lanka	0.09	2
Switzerland	0.09	2
Azerbaijan	0.05	1
Burkina Faso	0.05	1
China	0.05	1
Lesotho	0.05	1
South Africa	0.05	1
Tanzania	0.05	1

Source: Eleven data
Data collected between November 2012 and December 2015

For the same reference period, almost half of the supported startup teams were from Bulgaria, while a significant share of the funding went to startup teams from Serbia, Romania, Croatia and Slovenia, the remaining Balkan countries with stronger entrepreneurial ecosystems (see Table 4.11).

As regards team formation, half of the surveyed startups were comprised of team members who had known each other for three or more years prior to receiving finance. The long-term nature of the relationship between startup founders could be explained by at least two factors (see Fig. 4.1).

Firstly, the investment decisions taken during this time period were consciously based on a belief that the success of a startup team depended on well-consolidated personal relations between founders. Secondly, at this very early stage of development of the regional ecosystems, the startup communities were not that numerous, so their members resembled a close-knit network of professional and personal contacts rather than open-access communities driven by entrepreneurial spirit. Gradually, with the increase in the size of the entrepreneurial communities in the

Table 4.11 Stated headquarters of accepted ventures

Headquarters	Per cent (%)	Frequency
Bulgaria	49.31	71
Serbia	11.81	17
Romania	8.33	12
Slovenia	8.33	12
Croatia	4.17	6
USA	4.17	6
Germany	2.08	3
Macedonia	2.08	3
Montenegro	2.08	3
Armenia	1.39	2
Hungary	1.39	2
Spain	1.39	2
Argentina	0.69	1
India	0.69	1
Kyrgyzstan	0.69	1
Latvia	0.69	1
Sri Lanka	0.69	1

Source: Eleven data
Data collected between November 2012 and December 2015

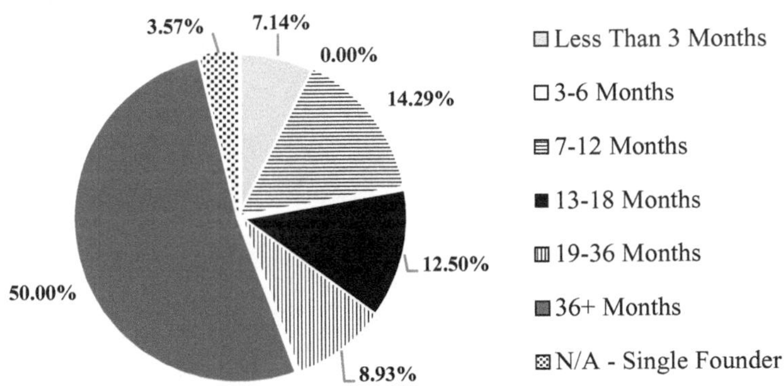

Fig. 4.1 For how long have team members known each other? Source: Startup Genome Survey Data, 2014, 56 startups Eleven Accelerator Venture Fund

region and the increase of the social capital and personal reputation that their members have accumulated, the long-term nature of the relationships becomes less prominent, yet it remains important as a determinant of investor confidence in new venture success. This is not surprising, given the fact that a good share of entrepreneurial ventures fails not so much because of an ill-identified market opportunity or mediocre value proposition but because of team dynamics.

The long-term relations between team members during the early years of the development of the entrepreneurial ecosystem are partly explained by the professional backgrounds of the founders, as they were mostly technically trained (engineers and software developers) and had known each other since university. In fact, the very first batches of startups were comprised of teams that consisted largely of technical members. Subsequent startup teams had a more balanced composition of technical and non-technical members (see Fig. 4.2). Whilst the experience of startup teams in more advanced entrepreneurial ecosystems is that teams that have a balance between technical and non-technical members do better, in the Balkans there has been a clear early bias in the financing of more wholly technical teams. This bias is now being overcome and later startup batches include fewer than 50% of technical member-dominated startups.

While the predominant focus of startup teams is customer development and customer acquisition, product development is almost equally as important (see Fig. 4.3). In some cases, a focus on engineering is justified,

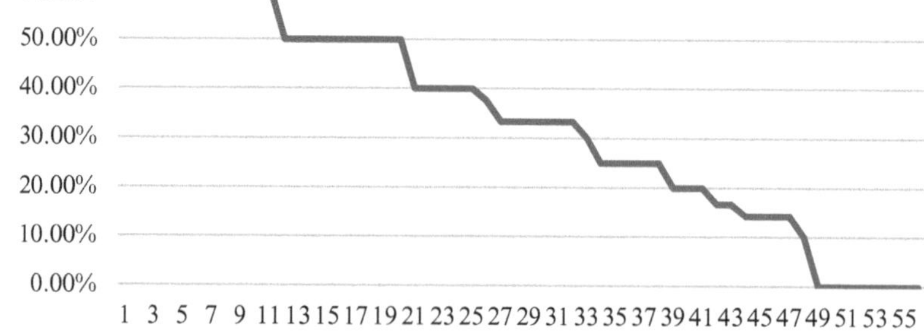

Fig. 4.2 Percentage of technical team members out of total team size. Source: 56 startups of the portfolio of Eleven

but in other cases, it arises from a myopic outlook about what the startup project needs, coupled with a lack of relevant marketing and business development skills amongst startup members. Figure 4.4 and the testimonials and case studies found in the second part of this book illustrate this aspect of the Balkan entrepreneurial dynamic more vividly.

In spite of the fact that our survey of Balkan entrepreneurship included early-stage ventures whose existence could only be justified only if they had a paying customer, a third of them focused on building the first version of the product, while only 22% were primarily involved with understanding

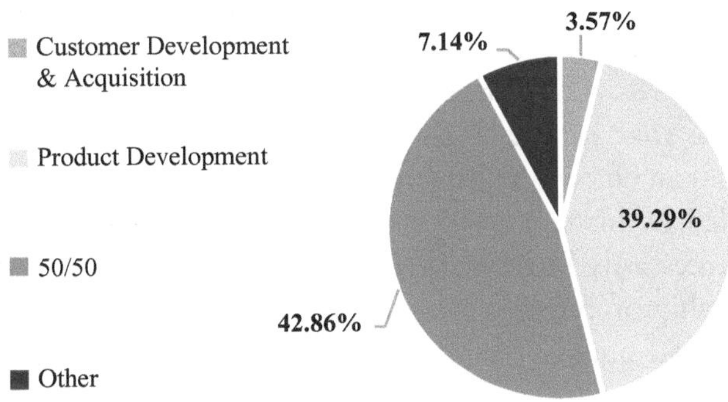

Fig. 4.3 Focus of the team. Source: 56 startups of the portfolio of Eleven

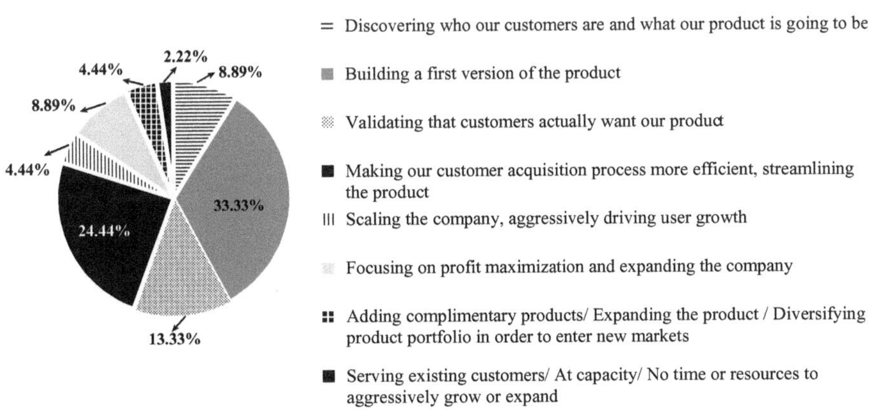

Fig. 4.4 Startups' current focus. Source: Startup Genome Survey Data, 2014, 56 startups Eleven Accelerator Venture Fund

who the customers were and whether they actually wanted the early venture's product. Rather than indicating a relative strength in understanding their customer, this behaviour shows that Balkan entrepreneurs still need to develop skills and an understanding of the factors that make startup ventures successful. Developing a customer-focused mindset is still an outstanding task for many early-stage ventures in the region. Because so many of these startups are geared towards foreign or even global markets, there is an urgent need for them to combine their engineering approach to entrepreneurship development with better customer understanding.

Not surprisingly, Balkan early-stage ventures report that their biggest challenges lie in customer acquisition, business model design, funding and product–market fit, in this order. Product development comes in at fifth place (see Fig. 4.5). Importantly, it is not financing nor a lack of technical skills that are the biggest challenges facing the development of successful Balkan entrepreneurial ventures; it is their business culture and skills, crippled by historical circumstances and cultural traditions. They are also unnecessarily burdened with overconfidence and an outlook on entrepreneurship of 'business is easy while engineering is hard' that needs to change. Good-quality business education seems to be an obvious fix

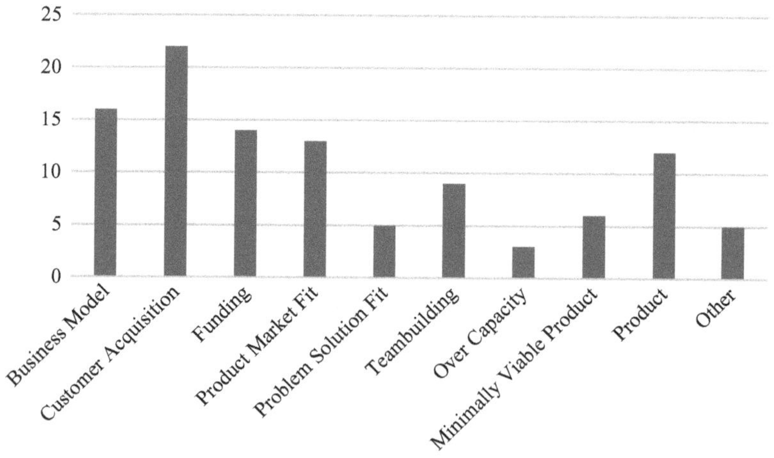

Fig. 4.5 Key challenges startup is facing today. Source: Startup Genome Survey Data, 2014, 56 startups Eleven Accelerator Venture Fund

for this situation, but this faces its own challenges, the strongest being that many members of the Balkan entrepreneurship community opine that business education in the region adds little or nothing to the probabilities of success of local ventures.

Most Balkan startups self-identify the source of their competitive advantage as team composition and technology (see Fig. 4.6). Both aspects are interrelated in many cases, as the long-lasting personal relationships amongst the team members of these first startup ventures were initiated during their technical training or professional career.

Despite the fact that regional entrepreneurial ecosystems are very young, Balkan entrepreneurial ventures are ambitious. Most startup founders perceive their products and services as genuinely global and see North America as their biggest current market, followed by Central and Eastern Europe (see Figs. 4.7 and 4.8). The next target market for business operations is either Western Europe or North America.

The expectation of the startup founders surveyed was that this accelerated international expansion in the richest international markets would happen with the help of a local partner and within the following year of operations. A rather optimistic scenario! (see Figs. 4.9 and 4.10).

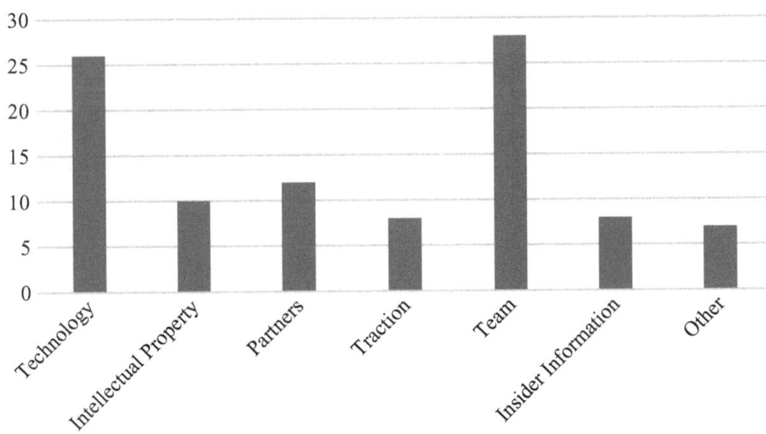

Fig. 4.6 Competitive advantage of the startup. Source: Startup Genome Survey Data, 2014, 56 startups Eleven Accelerator Venture Fund

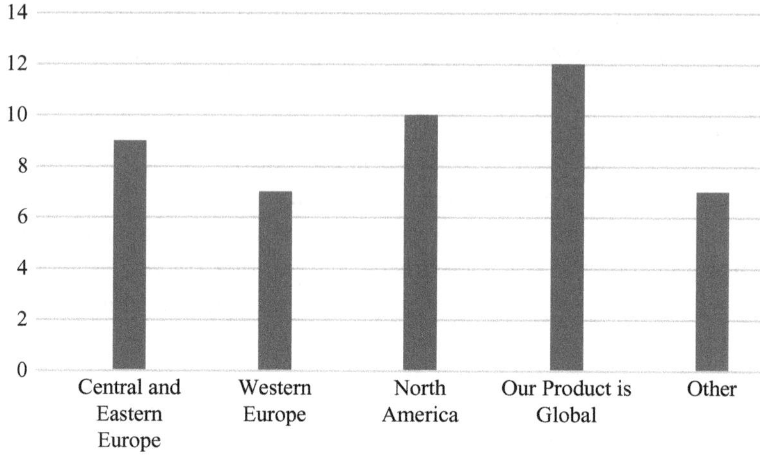

Fig. 4.7 Primary geographical market for product/service. Source: Startup Genome Survey Data, 2014, 56 startups Eleven Accelerator Venture Fund

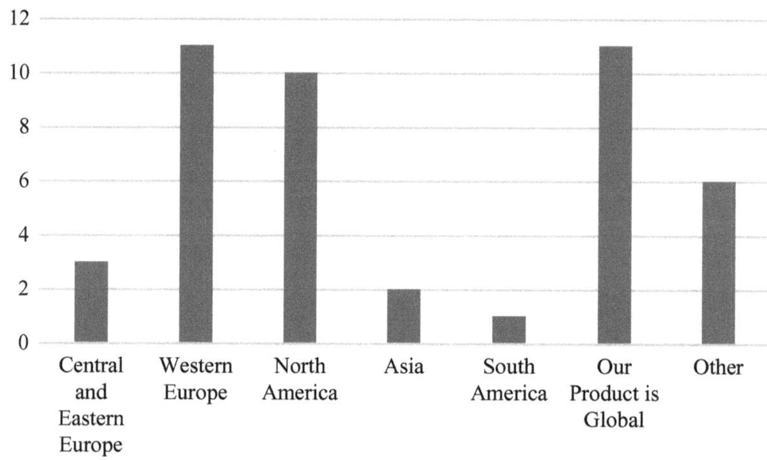

Fig. 4.8 Next geographical region considered for international expansion. Source: Startup Genome Survey Data, 2014, 56 startups Eleven Accelerator Venture Fund

Balkan entrepreneurial ventures were in their infancy in the period 2013–15. Most of the startups surveyed had already reached most milestones (or phases) during the Discovery stage, as reported in Fig. 4.11. During this stage, the most challenging milestones of those identified along the horizon-

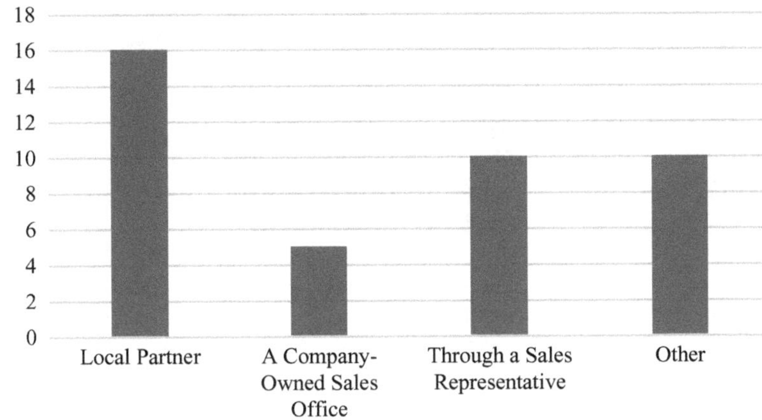

Fig. 4.9 How they plan to become active in the next geographical region. Source: Startup Genome Survey Data, 2014, 56 startups Eleven Accelerator Venture Fund

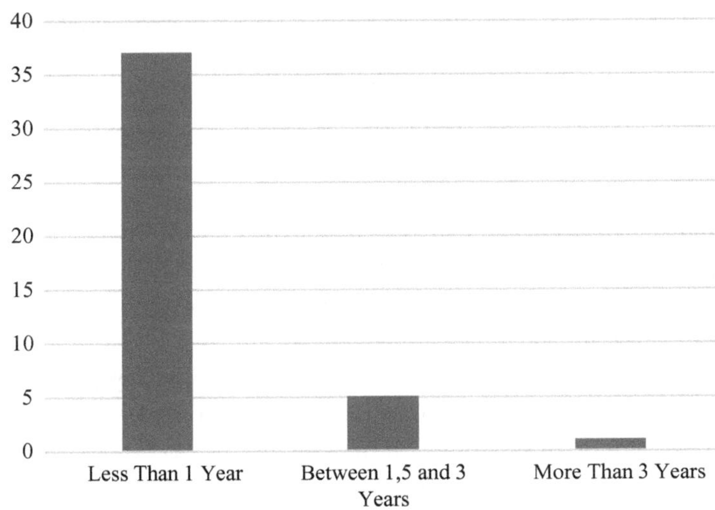

Fig. 4.10 Time expected to wait until they move to next geographical region. Source: Startup Genome Survey Data, 2014, 56 startups Eleven Accelerator Venture Fund

tal axis in Fig. 4.11 were: 'Many customer interviews are conducted' and 'Mentors come onboard'. Both reveal aspects of the Balkan entrepreneurial ecosystem that need to be strengthened substantially: customer and market knowledge as well as mentor availability and relevant experience.

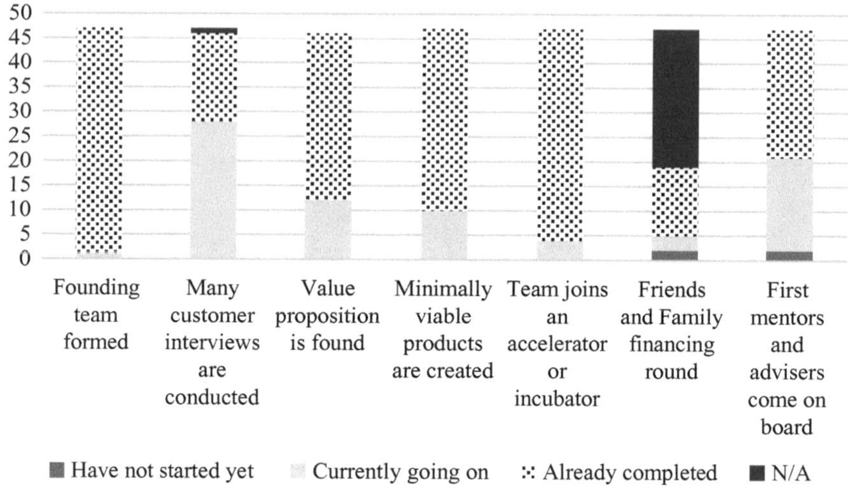

Fig. 4.11 Completion of development phases during Discovery stage. Source: Startup Genome Survey Data, 2014, 56 startups Eleven Accelerator Venture Fund

During the Validation stage, the most challenging milestones were customer acquisition and seed funding. At this stage, the startup ventures actively sought a product–market fit and pivots were not unusual (see Fig. 4.12).

During the Efficiency stage, Balkan startups felt the most prepared to overhaul the customer acquisition process and refine their value proposition. On the other hand, they felt their greatest challenges were achieving viral growth and establishing a repeatable sales process (see Fig. 4.13).

Finally, at the Scaling stage, only a few startups had worked towards milestone completion. Two milestones appeared to be more feasible: executive hire and back-end scalability (see Fig. 4.14)

The progress made by startups from stage to stage is related to their ability to learn in a consistent fashion. Startups that move consistently through the four stages (Discovery, Validation, Efficiency and Scale), advancing to the next one only after having achieved the milestones of the previous stage, are considered consistent. On the contrary, those that jump to the next stages without completing all the milestones of the previous stages are viewed as inconsistent. To measure this, we developed a proxy for inconsistent startups: those that were engaged in at least three activities that belonged to a later stage of startup development without

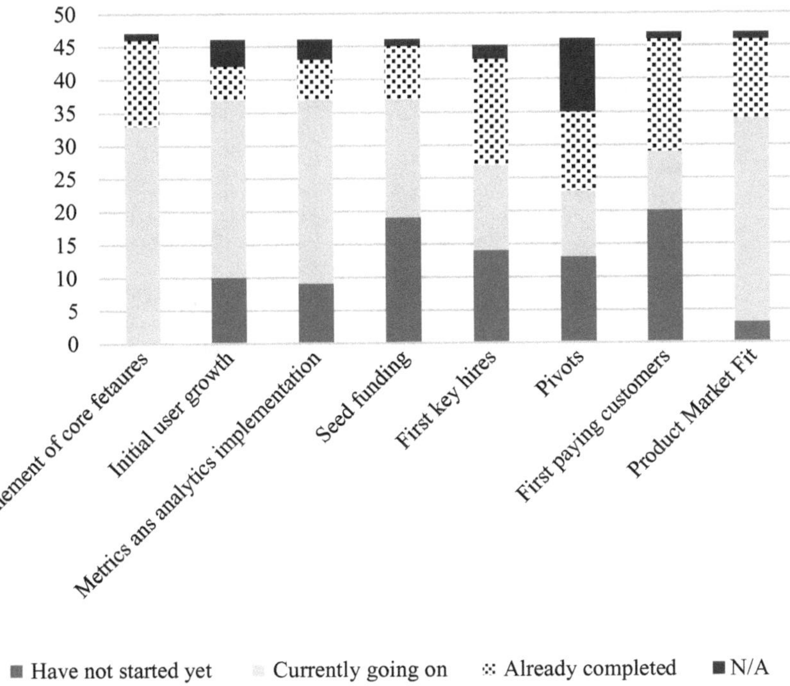

Fig. 4.12 Completion of development phases during Validation stage. Source: Startup Genome Survey Data, 2014, 56 startups Eleven Accelerator Venture Fund

having reached all the milestones of the previous stages. In Fig. 4.15, we document the proportion of consistent and inconsistent learners amongst the early-stage ventures we surveyed from the Balkan region.

There turned out to be more inconsistent than consistent startups. If the findings about the role of consistent startups are to be applied in the Balkan context, the learning processes of regional ventures can be improved by adding more discipline, direction and clearer priorities to the entrepreneuring process. On the other hand, at least if measured by their ability to hire part-time labour, inconsistent startups in the Balkans are not suffering from any evident disadvantages in their ability to attract talent (see Fig. 4.16). Once again, Balkan entrepreneurial ventures appear to offer a provocative counterexample to the general principle and invite us all to clearly identify the boundary conditions of commonly accepted conventions in entrepreneurship.

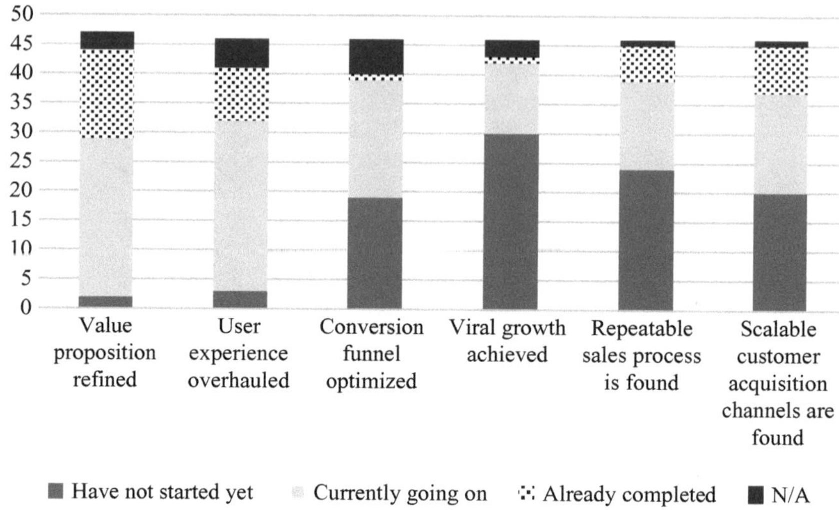

Fig. 4.13 Completion of development phases during Efficiency stage. Source: Startup Genome Survey Data, 2014, 56 startups Eleven Accelerator Venture Fund

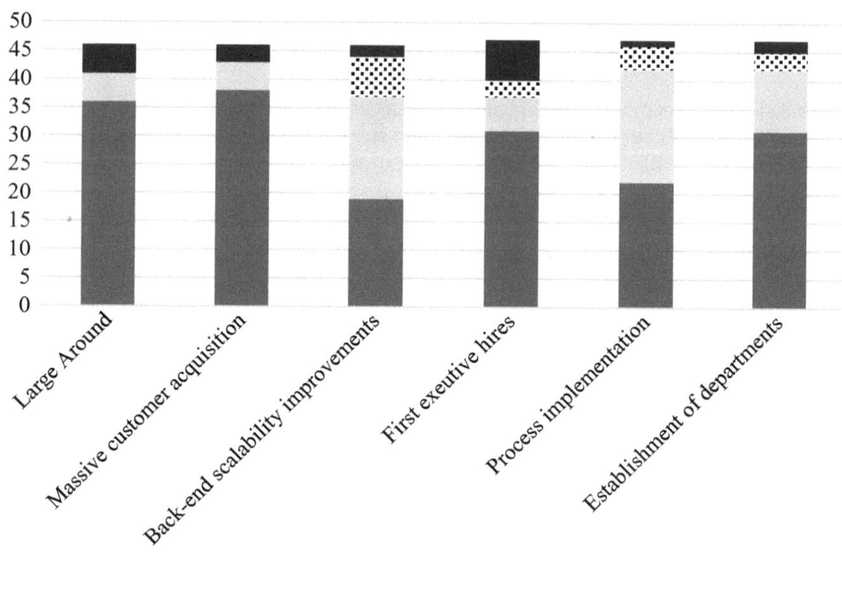

Fig. 4.14 Completion of development phases during Scale stage. Source: Startup Genome Survey Data, 2014, 56 startups Eleven Accelerator Venture Fund

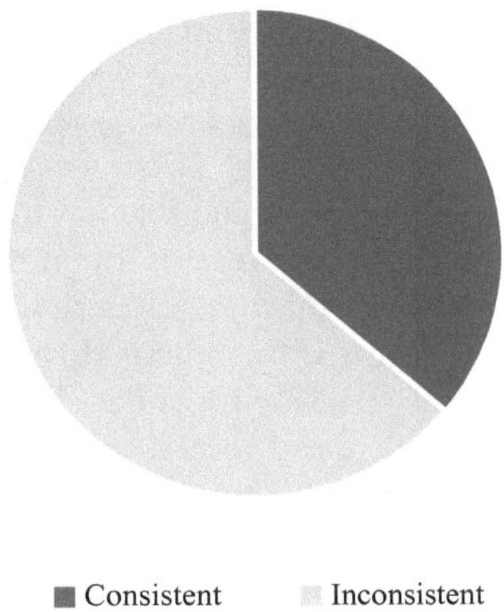

Fig. 4.15 Consistent vs. inconsistent startups. Source: Startup Genome Survey Data, 2014, 56 startups Eleven Accelerator Venture Fund

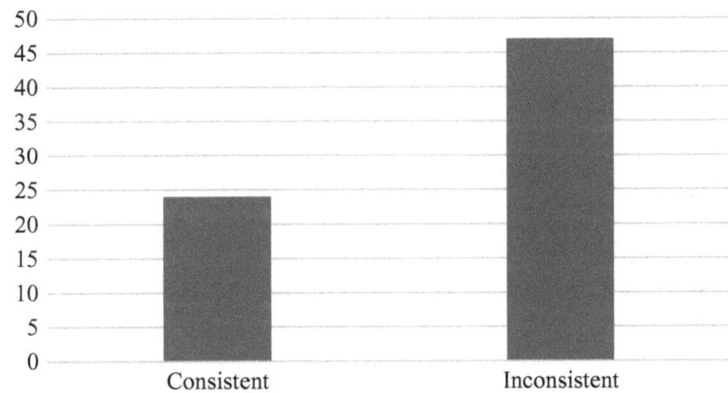

Fig. 4.16 Part-time employees in consistent and inconsistent startups. Source: Startup Genome Survey Data, 2014, 56 startups Eleven Accelerator Venture Fund

Developing Strategic Thinking in Entrepreneurs in the Balkans

Balkan entrepreneurial ventures are a fascinating object of study, as they can be seen as a microcosm of the region. Their technical talent is probably their most obvious strength and the lack of business 'savvy' and strategic thinking about business development and growth their most evident weakness. The latter must be rapidly developed in order to see Balkan entrepreneurial ecosystems consolidate into significant global centres for entrepreneurship rather than the source of exotic anecdotes.

Strategic thinking is a frame of mind that has five essential elements. It subscribes to a systems perspective; it is intent driven and characterized by intelligent opportunism; it requires thinking about the time dimension of decisions and it is hypothesis driven (Liedtka, 1998). A systems perspective involves the ability of decision makers to develop a mental model of the complete end-to-end system of value creation and understand the interdependencies within it. This entails the ability to understand the organizations and the context in which they are embedded in an integrated way. Strategic thinkers are intent driven and have a clear vision of the future. They are capable of mobilizing their energies and resources and, ignoring distractions, moving decisively towards their desired goal. Strategic thinkers show also ambidexterity, that is, they are capable of focusing on efficiency in many of the processes involved without forgoing an opportunity to explore new directions for organizational development. They can envision how processes evolve in time and embrace hypothesis testing as a method to uncover the true nature of things. Strategic thinkers are those individuals who will lead the region towards economic and social prosperity and, therefore, this research endeavour required an understanding of to what extent the mindsets of the key stakeholders in the Balkan entrepreneurial ecosystems possessed some or all these elements. We discovered that entrepreneurs in the region need to improve a great deal before they begin to resemble the type of strategic thinkers who might lead the Balkans towards significant improvements in prosperity and wellbeing, but we also found that Balkan entrepreneurs and their ventures have an enormous potential to become a significant positive force in these societies and economies. Balkan entrepreneurial ecosystems are the ideal place for this positive force to be nurtured and shaped, and by giving this account of their unquestionable achievements, we celebrate their incredible stories.

References

Bach, N., & Zubrinić, D. (2013). Mate Rimac young Croatian inventor created the fastest electric car in the world. *Croatian World Network*. Retrieved from http://www.croatia.org/crown/articles/10358/1/Mate-Rimac-young-Croatian-inventor-created-the-fastest-electric-car-in-the-world.html

Compass Magazine. (2018). Mate Rimac—Young Croatian inventor aims to electrify the entire world of transportation. *Compass Magazine*. Retrieved from https://compassmag.3ds.com/cn/11/Business/MATE-RIMAC

Crunchbase. (2018). Rockaway capital. *Crunchbase*. Retrieved from https://www.crunchbase.com/organization/rockaway-capital#section-overview

CTV News. (2016). *Meet the Croatian inventor behind the world's first electric supercar*. Retrieved from https://www.ctvnews.ca/autos/meet-the-croatian-inventor-behind-the-world-s-first-electric-supercar-1.2805660

Liedtka, J. (1998). Strategic thinking: Can it be taught? *Long Range Planning, 31*(1), 120–129.

O'Hear, S. (2016). Search startup Gjirafa scores $2M to bring Albania online. *TechCrunch*. Retrieved from https://techcrunch.com/2016/02/10/gjirafa/

Reback, G. (2016). Kosovo's Gjirafa search engine scores $2M as Google fails Albanian speakers. *Geektime*. Retrieved from https://www.geektime.com/2016/02/12/kosovos-gjirafa-search-engine-scores-2m-as-google-fails-albanian-speakers/

Rimac Automobili. (2018). *Technology*. Retrieved from http://www.rimac-automobili.com/en/technology/

Thompson, I. (2016). *Albanian*. About world languages. Retrieved from http://aboutworldlanguages.com/Albanian

Waldo, L. (2014). Meet the 2014 founders: Gjirafa, Albania/Kosovo's answer to Google. *Startup Yard*. Retrieved from http://startupyard.com/meet-the-2014-founders-gjirafa-albaniakosovos-answer-to-google/

5

Founder and Stakeholder Interests: Combining Economic and Societal Factors in the Balkan Peninsula

Introduction

Business models can be understood as cognitive schemas (Martins, Rindova, & Greenbaum, 2015), that is, 'images of real systems such as real business models that are shaped by the managers' own cognitive frames' (Massa, Tucci, & Afuah, 2017). Viewed from this perspective, there are two expected outcomes in the process of business model innovation. The first consists of a reconfiguration of a business model through the incorporation of a structure of activities and relationships that have been proved to have a value-creating effect in a different context. The second is a variation of an existing model that possesses some significant novel elements, which add value. These dynamics are expected to hold for all types of organizations, including early-stage ventures in the Balkans.

Early-stage ventures or startups are new firms that have not yet established effective work roles and relationships and that lack a track record of effective interactions with external stakeholders (Baum, Calabrese, & Silverman, 2000). At the earliest stages of venture creation, startup

This chapter is co-authored by Jonathan Pérez, School of Management, Universidad de los Andes, Bogotá, Colombia.

V. Andonova et al., *Entrepreneurial Ecosystems in Unexpected Places*,
https://doi.org/10.1007/978-3-319-98219-9_5

founders' identities, which are the ways they define what to do in relation to opportunities, are pivotal (Dutton & Dukerich, 1989). Technically, startup founder identities can be understood as the process of sensemaking of founders' actions on the bases of their intermediate set of beliefs (Albert & Whetten, 1985; Fiol & Dunbar, 1990; Fiol, 1991). Organizational members' deep identification with what they value and with the outcomes they wish to produce brings coherence to any organization (Fiol, 2001), including early-stage ventures. Therefore, the genesis of early-stage ventures cannot be understood unless the founders' sensemaking of what they do is taken into account simultaneously with the interests of investors, as these represent a reflection of a larger set of contextual norms.

Prior research has shown that startup founders are motivated by a mix of factors that combine their values and desirable outcomes, such as building great products, changing the world, creating a quick 'flip' and making a good living (Startup Genome, 2012). These motivations can be understood as belonging to a spectrum that ranges from the pursuit of one's own interests, relative success and dominance over others to values that emphasize concern for the interests and welfare of others (Schwartz, 2012). While concepts such as social entrepreneurship (Alvord, Brown, & Letts, 2004) and shared value (Porter & Kramer, 2011) have been proposed with the intention of clarifying the interactions between private and social outcomes, the boundaries of these theoretical frameworks are still fuzzy and in the process of construction. What seems to be generally accepted is that it is not evident how to reconcile private interest and societal purpose and that innovative solutions and entrepreneurial undertakings are instrumental in achieving this integration (Haugh, 2005). The associated challenge of harmonizing private economic and societal purpose can arguably be achieved through business model innovation and mirrors a more general structure of societal value relations, that of personal versus societal focus (Schwartz, 2012).

Against this backdrop, we show that stakeholders from the market environment powerfully shape the business model innovation process of early-stage Balkan ventures that pursue both private economic and societal purpose. Following Albert and Whetten (1985), we observe that the process of business model innovation in early-stage ventures is incorporated through either substitution or addition of new behaviours provided by the environment, and is adaptive as a result. Thus, business model innovation is conceived as a process in which founders' identities interact with market stakeholder interests and morph along the lines of personal versus societal focus, to achieve a viable product–market fit. Eventually, startups persist and business models emerge only if they are consistent with the identities of the founders and the interests of market stakeholders.

The long-term viability and potential impact that organizations with private economic and social purpose can bring cannot be well understood without a deep understanding of the genesis of these ventures. The purpose of this chapter is to identify and describe the genesis of startup ventures that reconcile private economic and societal purpose. How do their business models come into existence? How do market stakeholders, such as return-maximizing investors and utility-maximizing users, shape the product–market fit of these startups and hence the balance between economic and societal purpose? In addressing these questions, we expose the dynamic interaction between founders' sensemaking process and market stakeholder interests that give rise to early-stage ventures in the context of the Balkans. The data for this chapter comes from the investment portfolio of Eleven, the accelerator venture fund located in Bulgaria. Eleven was approved as fund manager of the JEREMIE, a European Investment Fund accelerator and seed fund instrument, in early 2012. Eleven held 10 application windows during its investment period, which ended in 2016, receiving about 3500 applications and approving 116 investments, mostly from Southeastern Europe (Eleven quarterly report, 2017).

The Genesis of Early-Stage Ventures

Entrepreneuring Profit and Purpose

At a general level, 'entrepreneuring' is defined as 'efforts to bring about new economic, social, institutional, and cultural environments through the actions of an individual or a group of individuals' (Rindova, Barry, & Ketchen, 2009). This all-encompassing definition affirms a multidimensional concept of purpose that is consistent with different kinds of entrepreneurial activity. From there, the concept of entrepreneuring has given rise to a broad range of categorizations and taxonomies, including innovation-driven enterprises versus SMEs (Aulet & Murray, 2013), necessity-driven versus opportunity-driven versus improvement-driven entrepreneurship (GEM, 2017) and social entrepreneurship (Short, Moss, & Lumpkin, 2009). Besides the fact that these taxonomies were created in order to provide a common understanding of a complex phenomenon, there has generally been little consensus about how exactly to define these domains of entrepreneurship.

The innovation-driven entrepreneuring approach, for example, emphasizes the pursuit of global opportunities with high growth potential based on innovations that give rise to competitive advantage. These innovations are expected to come in many varieties, including technological innovations, business model innovations, process innovations and others (Aulet & Murray, 2013). The emphasis here is on rapid and exponential growth that is possible thanks to the scalability of the resources that forms the basis of the competitive advantage. Thus, the prospects of innovation-driven enterprises are radically different from the prospects of SMEs, and innovative entrepreneurship has been considered the engine of modern economic and social development by facilitating the reallocation of resources from less to more productive uses, arguably improving societal wellbeing (Acs & Storey, 2004; Gries & Naudé, 2010; Hausmann & Rodrik, 2003; Minniti & Naudé, 2010).

The necessity-driven versus opportunity-driven debate evolves along similar lines of interaction between entrepreneuring and social and economic contexts. Necessity-driven entrepreneurs are those who are forced into entrepreneurship because other work opportunities are not available or are unsatisfactory (Aidis, Welter, Smallbone, & Isakova, 2007; Bosma & Harding, 2007; Harding et al., 2006; Minniti, Bygrave, & Autio, 2006; Smallbone & Welter, 2004), while opportunity-driven entrepreneurs are those who have been attracted to entrepreneurship by the desire for self-realization. Among the opportunity-driven entrepreneurs, those who are improvement-driven are those who seek to either earn more money or be more independent. While these categories have been part of the Global Entrepreneurship Monitor survey, the dominant data source of the degree and nature of entrepreneurship across countries, there is an active debate calling for a more nuanced approach, putting centre-stage the need for a discussion about the meaning of entrepreneurial opportunity (Casson & Wadeson, 2007; Williams, 2008). In this debate, the concept of entrepreneurial opportunity is also analysed from either a private commercial (Davidsson, 1991) or a social perspective (Hockerts, K. 2006).

Social entrepreneurship is at the centre of these arguments. Most definitions of social entrepreneurship deal with specific dimensions that categorize this activity (Mair, Battilana, & Cardenas, 2012) and highlight aspects such as legal status (for-profit, not-for-profit and hybrid), experimentation with organizational forms (alliances, partnerships and joint ventures), defining the ultimate objective as the triggering of a 'catalytic or systemic change in the context' (Mair et al., 2012) and producing a change in the social systems not just as a byproduct (Mair et al., 2012). Mair et al. (2012) propose a systematization of social entrepreneurship models that leverages different types of resources and capital to produce social change. They document the existence of four dominant models based on political, human, economic and social capital. Of these, it appears that the entrepreneurial model that leverages economic capital and combines market principles deserves special attention, as there is an increasing number of social entrepreneurs creating their ventures to combine a social mission with commercial activities (Battilana, Lee, Walker, & Dorsey, 2012; Battilana & Dorado, 2010). Yet, from the position of social entrepreneurship, in the union between social purpose and private economic interest, there exists a real risk of prioritizing profit seeking over social mission.

Others have taken the opposite view and have put the emphasis on the investors' rationale behind entrepreneuring. According to this stream of mostly practice-oriented research, the capital asset pricing model (CAPM) is still the dominant paradigm regarding investment decisions and, therefore, entrepreneurial decisions should be primarily based on the well-known financial return on investment (Grabenwarter & Liechtenstein, 2011). The concept of return dominates according to this view and 'any business model in which every business unit of social/environmental impact has a cost in terms of financial return is therefore inevitably a disguised form of philanthropy' (Grabenwarter & Liechtenstein, 2011).

In this domain, investment strategies whose objectives expand beyond purely financial return are classified in several groups, among which the most common are corporate social responsibility investing, responsible investment based on ESG (environmental, social and governance) standards and impact investments that strive to integrate financial returns with objectives for sustainable development. From the perspective of the latter, the concept of a trade-off between financial return and impact does not reflect market reality and is seen as 'meaningless for the purpose of characterizing the investment space' (Grabenwarter & Liechtenstein, 2011).

Because we focus on the genesis of startup ventures that have both a private economic objective and a social purpose, our goal is to uncover how business models' characteristics combine a change in the social system (Alvord et al., 2004) and economic profit generation. We treat the presence of a verifiable social purpose as a sufficient condition to treat early-stage new ventures as organizations that pursue a social mission. By combining extant views, we suggest that both aspects are necessary and indispensable: the socially conscious entrepreneurial motivations of startup founders and the investors' rationale of maximizing return on investment are essential elements in the business model innovation process. The central question, then, is how these business models come into existence.

Business Models and Business Model Innovation

Extant literature contains a spectrum of definitions of business models because researchers understand business models differently (Zott, Amit, & Massa, 2011): an architecture, including a description of its components (Timmers, 1998); a system of interdependent activities (Zott & Amit, 2010); a heuristic logic (Chesbrough & Rosenbloom, 2002); stories (Magretta, 2002); a cognitive structure (Doz & Kosonen, 2010); a representation of interrelated variables (Morris, Schindehutte, & Allen, 2005); a reflection of the firm's realized strategy (Casadesus-Masanell & Ricart, 2010). Despite the broad range of views on business models, some researchers have identified a convenient convergence in recognizing business models as a 'design or architecture of the value creation, delivery, and capture mechanisms' of a firm (Teece, 2010). Massa et al. (2017) systematize masterfully different views and cluster the proposed interpretations and function of business models in three broad domains: (i) business models as attributes of real firms, (ii) business models as cognitive/linguistic schemas and (iii) business models as formal conceptual representations of how a business functions.

Scholars who view business models as attributes of existing firms argue that these attributes are not observable but still permeate firms' activities, according to which business models can be classified. Some of the activities that form part of the outcomes of a business model according to extant research include money making (Birkinshaw & Goddard, 2009); mapping purpose and value-creation opportunities (Bocken, Rana, & Short, 2015); revealing firm strategy (Casadesus-Masanell & Zhu, 2010); formulation of competitive strategy (Chesbrough, 2010); description of interrelations and connections for value creation (Nielsen & Lund, 2014); facilitating value creation and capture (Smith, Binns, & Tushman, 2010); and representation of the foundations for the focal firm's value capture (Zott & Amit, 2010). Scholars within this research stream also study outcomes associated with business model attributes such as activities, resources and capabilities and underlying core logics and strategic choices (Dahan, Doh, Oetzel, & Yaziji, 2010).

According to a different view, business models can be seen as formal conceptual representations of how a business functions (Massa et al., 2017). These formal conceptual representations are explicit, written down in pictorial, mathematical or symbolic models that provide simplifications of real systems. Some researchers consider business models to be translations or representations of relevant aspects of the organization's activities, which may vary in the level of abstraction, content and semantics. Semantic variations have created a lack of agreement on what the critical parts of a business models are and a confusing absence of consistency in the use of the terminology surrounding them. Some of the functions of the business models according to this interpretation of business models as conceptual representations of how business works are the representation of the firm's money-earning logic (Abdelkafi & Täuscher, 2016); the specification of how a new venture can become profitable

(Boons & Lüdeke-Freund, 2013); the illustration of the production/delivery system (Itami & Nishino, 2010); the explanation of the design or architecture of the company's mechanisms to create, deliver and capture value (Reim, Parida, & Örtqvist, 2015); and the description of how a firm interacts with the surrounding ecosystem (Wells, 2016).

Lastly, business models are viewed as cognitive or linguistic schemas. According to this view, business models are defined as images of real systems and much of the research is focused in how the organization's members make sense of the business model and the role of this sensemaking in social interactions. The most influential constructs within this interpretation are business models as managers' perceived logic of how the business unit functions and creates value (Aspara, Lamberg, Laukia, & Tikkanen, 2013), a heuristic logic (Chesbrough & Rosenbloom, 2002), cognitive structures (Doz & Kosonen, 2010), stories (Magretta, 2002) and reflections of managers' mental schemas. These basic constructs of business models are usually paired with some key functions, such as the connection between the firm and the market environment (Aspara et al., 2013); sensemaking of the environment and the organization; and the relationship between the firm's identity and the external world (Gioia, Schultz, & Corley, 2000).

In studying the genesis of business models in a transforming economy, we take the construct of business models as cognitive schema. From this perspective, the process of innovation and reconfiguration of business models occurs in a number of ways. Martins et al. (2015) emphasize the analogical reasoning and the conceptual combination that are facilitated by the interactions of the founders with other actors in the ecosystem. Berends, Smits, Reymen and Podoynitsyna (2016) argue that experiential learning and cognitive search are part of the business model's innovation trajectories, while Sosna, Trevinyo-Rodríguez and Velamuri (2010) propose that in the case of early-stage ventures, when the business model is still being conceptualized, trial and error are instrumental for business model development and validation by market realities. Thus, business models seen as 'images of real systems—such as real business models—that are shaped by managers' [founders'] own cognitive frames' (Massa et al., 2017) are moulded by both the founders' identities and the interests of key stakeholders, notably of those who belong to the market domain.

The cognitive view emphasizes the identity and mental schemas of founders as well as the reconfigurations and metamorphosis of business models that emerge from the shifting identities. A number of constituencies are essential for uncovering the business model innovation process for early-stage ventures with societal purpose: socially conscious startup founders, return-minded investors and utility-maximizing consumers.

Values and Entrepreneurship: Startup Founders' Identities and Investors' Interests

Identity and sensemaking are intrinsically linked to personal values, and these are instrumental for understanding social organization and change, including entrepreneuring. As Schwartz (2012: 3) puts it, 'When we think of our values, we think of what is important to us in life'. Values are a cohesive force within personal identity and, as such, values are the primary phenomenon in the experience of personal identity (Hitlin, 2003). Consequently, values lead to the development of 'value-identities', which are phenomena that socially situate individuals with respect to others.

Empirical research suggests that values can enhance or pose barriers to entrepreneurial activity (Davidsson, 1995; Davidsson & Wiklund, 1997; Kreiser, Marino, Dickson, & Weaver, 2010; Schmutzler, Andonova, & Diaz-Serrano, 2018; Thomas & Mueller, 2000). For example, individualism/collectivism is one of the most extensively studied value dimensions in the entrepreneurial domain. Individualism is associated with the notion that individuals seek and pursue personal goals and that the individual's needs take precedence over the needs of the group (Hofstede, 1980; House, Hanges, Javidan, Dorfman, & Gupta, 2004; McGrath, MacMillan, & Scheinberg, 1992; Schwartz, 1994).

The notion that individualism favours entrepreneurship is based on the argument that the achievement orientation and the pursuit of personal goals, as values associated with individualism, are likely to favour entrepreneurship. The expected positive correlation between individualism and entrepreneurship has been confirmed empirically (McGrath et al., 1992; Thomas & Mueller, 2000; Shane, 1992; Wennekers, Uhlaner, & Thurik, 2002). On the other hand, it is argued that collectivistic values do not necessarily discourage entrepreneurship and, rather, values such as universalism and benevolence that are related to self-transcendence may lead to different motivations and forms of entrepreneurship. Erez and Earley (1993), for example, point out that Israeli entrepreneurs satisfy their need for affiliation through the creation of new ventures. Balkan countries are markedly collectivistic and, given their Communist past, this cultural characteristic is frequently seen as a hindrance to entrepreneurship. However, this chapter contains the first serious research endeavour of the ability of Balkan entrepreneurs to build viable business models that combine a well-established societal purpose, arguably triggered by a combination of the marked collectivistic orientation of these countries and economic logic.

In essence, values are expected to affect the type of entrepreneurship and the business models associated with entrepreneuring rather than the level of entrepreneurial activity. As one of the cardinal principles for organizing values, the personal versus societal focus is expected to profoundly affect entrepreneuring, and we argue that there are at least two immediate channels for this type of mindset—Balkan startup founders' identities and investors' interests.

Founders' Identities and Sensemaking

The concept of identity is used in psychology and sociology to describe an individual in relation to a larger cultural system in which they are embedded (Goffman, 1967). Within the literature on organizations, Albert and Whetten (1985) define identity as what is central, enduring and distinctive in an organization's character. An organization's identity can be defined by its response to the question 'Who are we as an organization?' and serves as a reference for members when they are interacting on behalf of their organization (Whetten, 2006). This definition is not necessarily derived from the organization as a whole but, usually, from each department, work team, division, subunit and so on (Ashforth & Mael, 1989). Whetten (2006) has built on this notion of organizational identity and has defined it as 'attributes used by an organization to positively distinguish itself from others'. Organizational and personal identities are expected to show very high level of correlation in the case of early-stage ventures.

At a personal level, identities stand for a set of beliefs through which organizational members make sense of their actions (Albert & Whetten, 1985; Dutton & Dukerich, 1989; Fiol & Dunbar, 1990). These identities explain how individuals understand what they do, how they behave and, ultimately, how they make sense of what the organization is (Fiol, 1991). Fiol (1991) argues that although an organization has a general system of rules that gives meaning to its behaviour, there is not a direct one-to-one link between these behaviours and their meanings. She suggests that individuals' organizational identity gives the context to attach behaviours to the rules that give them meaning. As a result, when we refer to individuals' identity within a larger system, we are alluding to people's understanding of themselves in relation to the broader system, and that eventually represents what they consider to be central, enduring and distinctive about their organization. It is essential to note that these perceptions will vary depending on 'where [people] sit' in the organization (Dutton and Jackson, 1989).

In the case of early-stage ventures, it is the founders' cognition and sensemaking that provide the most important input for their business model design (Kimberly, 1979; Sosna et al., 2010). Research on the effects of founders' identity on startup ventures suggests that these effects should be relatively strong, either because most decisions are made by one person (the founder) or because emerging firms are typically small entities that are in the process of being created by a small number of decision makers (Barney et al. 1998; Whetten & Mackey, 2002). In these circumstances, founders' identities affect critical characteristics of an emerging organization and its business model. A typology of founder identities highlights these effects (Fauchart & Gruber, 2011): Darwinian identities are focused more on winning the competition against other firms and are driven primarily by economic self-interest, revealing a strong individualistic orientation. 'Communitarians' view their firms as social instruments that uphold and are supported by a particular community, giving importance to a mutually beneficial relationship. 'Missionaries', on the other hand, see their firms as political objects designed to support a particular cause for the benefit of a society. Hybrid identity emerges through combinations of the three basic categories that in essence define the continuum between personal- and societal-oriented new ventures.

Eventually, both founders' identities and the associated startups' business models are subject to evolution and change. In a revisiting of her work, Fiol (2001) proposes that an organization with different and changing identities must have the adaptive capabilities to continuously regain advantages in the context of market competition. She refers to identity changes as those that might occur through identity additions and substitutions, as proposed by Albert and Whetten's (1985) early work. As such, changing identities not only could shift the nature of an organization's culture, in parallel to the change in the cognitive schema of the founders and members of the organization, but they alter the way these make sense of the business model in a specific context. We see the changes in the sensemaking process of founders of early-stage startups with societal purpose as instrumental for an understanding of the business model innovation process of these ventures.

Market Stakeholder Interests

Taken broadly, new ventures are leveraged by a complex system of interdependent elements, referred to as an entrepreneurial ecosystem, as described in Chap. 1 of this book. At the most intuitive level, entrepreneurial ecosystems consist of government policies, access to finance, culture, infrastructure and other support mechanisms, human capital and market-based interactions (Isenberg, 2011; Neumeyer & Corbett, 2017). The latter are particularly important for startups as their efforts are primarily directed to simultaneously achieving a product–market fit, that is, a situation where the product matches what customers in a specific market are interested in buying; and attracting finance, by convincing venture capitalists, angel investors and others to stake their capital on the startup's undertaking. Thus, at the very early stages of venture creation, startup founders are hugely dependent on stakeholders from the market environment as they are the key contributors to the business model innovation process.

We focus on the role of investors as market stakeholders because they, being powerful influencers, trigger adaptations in the business model and in this way also alter the underlying logics of justification (Mair et al., 2012) and identity of startup ventures. When it comes to ventures with a societal mission, however, more often than not the group of stakeholders includes foundations, private donors, public authorities, banks or social impact investors and excludes venture capitalists. In fact, in the investment literature, there is a clear differentiation between 'market-based' and 'mission-based' investing, implicitly suggesting that there exists a trade-off between financial returns and the social benefits investors create (Monitor Institute, 2009). As a result, returns-driven investors associate social mission with financial losses, while investors who seek social and environmental ends fear that a focus on profitability might undermine the noble cause of their investments (Grabenwarter & Liechtenstein, 2011).

Market stakeholders, and venture capitalists in particular, are expected to make rational decisions. This implies that they prefer more to less, a behaviour that has triggered wide debate about how economic activity at large can be made compatible with the sustainable development of society. However, even outside the context of these debates, 'it is hard to argue that any provider of capital would fund a newly created company purely for financial return considerations' (Grabenwarter & Liechtenstein, 2011). New ventures are normally financed on the basis of a large number of considerations, the most important of which is the choice of a business model that is attractive to and convincing for investors (Grabenwarter & Liechtenstein, 2011). As a result, from the point of view of venture capitalists and similar investors, startups that pursue a social mission represent nothing less than returns-driven investments that have an additional value-adding element. If they were not, the investments would be 'a disguised form of philanthropy' and would suffer from disadvantages in the scale of operation and limited prospects for growth (Grabenwarter & Liechtenstein, 2011).

Context, Data and Methods

During the second half of the twentieth century, Bulgaria and large parts of Central and Eastern Europe were ruled by a particular style of collectivistic ideology that gave rise to the so-called Communist regimes, a type of dictatorship that relied on planned economies. These regimes entailed the imprisonment of the economic and intellectual elites as well as the collectivization of private property. The fall of the Communist regimes in the 1990s saw the creation of institutions for private property protection and, in many cases, the restitution of confiscated assets. The ideology of collectivity was rapidly substituted by the idea of the supremacy of individual rights, which sometimes appeared as an irrational backlash against the previous social order.

Nevertheless, economic and social progress requires a certain—and frequently high—level of collectivity and societal purpose. The current rise of the entrepreneurial ecosystem in Bulgaria and other Central and East European countries is seen here as a restoration and legitimation of the need for collective and coordinated economic and social interactions but on a different basis, that of individual freedom and entrepreneurial ambition. As such, Bulgaria from 2000 onwards, when the local entrepreneurial ecosystem started to develop, represents a laboratory for understanding how early-stage entrepreneurial ventures with a societal purpose come into being out of a historical period characterized by the delegitimation of collectivity and a strong focus on individual rights and freedoms.

We conducted purposeful sampling to ensure that methodological rigour. First, we defined the selection criteria as follows: (i) founders of early-stage ventures who participated in Eleven's acceleration programme and received capital as a result of a congruence between the investors' objectives and the founders' identity; (ii) independently, early-stage ventures identified by investors as having an actual social/environmental purpose after the initial investment stage; (iii) cases where the majority of the investors agreed on the existence of a social/environmental purpose of the early-stage venture.

Second, in March 2016, we designed a structured interview template and a questionnaire and asked each of the three investors to reveal their investment philosophy. For each of the 116 ventures, each investor was asked independently to identify the existence of a clear social and/or environmental purpose in each of the listed startups, assigning a value of 1 to a venture with such a purpose. Then, the three investors scored the startups 0 (venture lacks a social/environmental purpose) or 1 (venture has a social/environmental purpose).

Third, we calculated the score for each startup as the sum of the values given independently by all investors. For our study, we selected the startups that met the criterion of a minimum score of 2, meaning that in these cases the social/environmental purpose of the venture was evident to more than one investor. Only 18 of the 116 early-stage ventures received a score of 2 or higher. Investor 1 identified 27 such ventures, Investor 2 identified 24 and investor 3 identified nine. The correlation coefficients between the investors' classifications ranged between 33% and 47%.

As of March 2016, when the interviews with investors were conducted, 14 of the 18 early-stage ventures were still functioning, while 4 were write-offs. Between March 2016 and October 2017, we contacted the founders of the remaining startups to conduct and record an interview. We obtained a positive response from seven of the ventures and we present this data briefly in Table 5.1. This translates into a response rate of 50%. Given that between six and eight participants are recommended as appropriate for a typical phenomenological study (McCracken, 1988; Smith & Eatough, 2006), we designated the founders of these ventures as our final sample.

Table 5.1 Startups with a recognizable social and/or environmental purpose

	Recognizable social and/or environmental purpose			
Startup name	Investor 1	Investor 2	Investor 3	Startup activity
Enhancv	Yes	Yes	No	A professional resume and CV builder
Escreo	Yes	No	Yes	Dry-erase paint producer
Farmhopping	Yes	Yes	Yes	An online platform that connects farmers with city dwellers
mBrainTrain	Yes	No	Yes	Producer of mobile EEG devices
Melissa Climate	No	Yes	Yes	Producer of solutions for climate control
Playground Energy	Yes	Yes	Yes	Producer of playground equipment for fun, health and education
Strawberry Energy	Yes	No	Yes	Producer of smart city solar benches

In October 2017, we added another startup case to our final sample, Grajdanite, an app that allows citizens to report dangerous behaviour on the road. The founders had previously applied to join Eleven's acceleration programme but had not received funding, reportedly because at the time there had not been a match between the objectives of investors, the founders' societal purpose and the startup identity. Eventually, towards the end of the fund's investment window, this venture received financing from Eleven, arguably in response to the adaptation and re-balancing of the societal mission associated with the founders' identities and the investors' interests.

The investors' interests and philosophy are captured in interviews with the three managing partners/investors of Eleven. It is evident that the dominant rationalization of investment decisions obeys a return-maximizing logic:

> I do not understand social investing and I do not believe in that. Whatever you do, you do it to solve a problem or address a need. The better you solve the problem the more value you generate. (…) Every good business has impact on society because it has a mission. This makes it better for society.—Investor 1, Eleven
>
> An investor will not invest in anything that harms any of these [social and environmental] aspects. The positive impact, however, has to be sustainable. It is very risky to start projects that are not well funded, so the social element is a feature. (…) If you have 2 startups that do exactly the same but one also has some social value then as an investor you invest in this one because it creates more value and this will help it succeed. If it succeeds than the ultimate goal is achieved. The [social or environmental] impact part of a business model is not the feature on which the investment decision is made. There are startups in which we did not invest because there was no one to pay the bill. This is where it falls down for me as an investor. If it does not have this mission but it is not damaging any of these aspects then it is OK from the investment perspective.—Investor 2, Eleven
>
> Ventures with social impact can be for profit. There is a lot of innovation that can have a deep social impact and that could be profitable. (…) We try to focus on the underlying business and the social element is a spillover. Just by being the best and improving some processes companies create social value. (…) We never specifically focused on social ventures. We want to invest in them as long as they have a reasonable business model.—Investor 3, Eleven

Highlighting the sensemaking process of venture founders in relation to their interactions with their market stakeholders is a matter of exploring the experiences involved in these processes, as perceived and accounted for by the founders.

We conducted semi-structured interviews in order to gain a deep understanding of the ways in which founders identified with the purpose of their new venture; the barriers and/or drivers they recalled of their interactions with the market stakeholders and their responses to that. The phenomenological interviews were designed to obtain a first-person description of some specific experience or events, where the participant established the line of the conversation (Cope, 2005). Therefore, we arranged the interviews in three parts: the first one enquired about their venture and how they self-identified with the purpose of it, letting the participants make their points and asking follow-up questions to zoom in on some of their statements. The second part of the interview contained questions related to their participation in the acceleration programme developed by Eleven, letting them lead the conversation and asking additional questions when they discussed the funding experience. The third part discussed the specific social/environmental value of the startup, again letting them lead the conversation and zooming in on aspects related to possible tensions they encountered regarding their specific societal mission and revenue model.

In line with the phenomenological approach and recommendations for the process of interpretative phenomenological analysis (Hycner, 1985), we studied the interviewing data in the following sequence: (i) gaining insights from the data, (ii) sensemaking of the experiences, (iii) categorization of the findings, (iv) pattern recognition, (v) interpretation of the identified patterns and lastly (vi) building bridges with extant theory. These six steps have previously been used in a similar way to analyse different experiences within the framework of interpretative phenomenological analysis.

We listened to the interviews and transcribed and read them to gain an understanding of the story as a whole. At the sensemaking stage, significant fragments of the interviews were highlighted, and we assigned them a unit of meaning, known as a code. Once the codes were identified, we clustered and categorized them according to themes that were relevant to the process, allowing us to construe a 'master-theme list' (Smith, Jarman, & Osborn, 1999). The pattern recognition step involved a comparison of themes across all the transcripts to find shared aspects of experiences, creating second-order categories that formed clusters of themes. The interpretation stage let us create a narrative that involved our interpretations of the shared experiences and the participants' account of those experiences in their own words, maintaining an inductive process and letting the data 'speaks for itself' (Smith et al., 1999). The final step allowed us to connect the findings with relevant literature in an iterative and comparative process that gave significance to the unique experiences of individuals while connecting them with extant theory, providing explanations and tracing the path to potential theoretical advances.

Case Study Analysis

In this section, we give an account of the genesis of the early-stage ventures identified in our research that reconcile private economic and societal purpose in their business model. We were able to identify five stages in the process of business model innovation, during which the founders brought a new venture into existence. In this process, both market stakeholders, that is, return-maximizing investors and utility-maximizing users/buyers, shape the product–market fit of these startups and consequently shape the balance between economic and societal purpose, contributing to the business model innovation process.

Table 5.2 shows the categories that were found to have the highest frequencies in our interviews. We report in descending order the categories that were referenced more than ten times. Together, the categories shown in Table 5.2 represent more than 63% of all identified themes mentioned by founders.

In Table 5.3, we show the second-order categories, which were derived from the information extracted from the founders' interviews organized in relation to the timing of the startup creation. We show the timeline of the venture creation process, the main categories and their relation to the supra-categories. All the other themes identified also fit into this frame and help construe the meanings of each second-order category.

Table 5.2 Categories, frequency and meaning

Category	Frequency	Meaning
Purpose	27	Founders talk about the social and/or environmental purpose as a motivator.
Teamwork	20	Founders talk about the importance of their team and the skills of team members.
Infrastructure	20	Founders refer to the digital infrastructure and labour market and the role of these conditions.
Legitimation	17	Founders talk about the acceptance of the value proposition by other market participants.
Experience	16	Founders discuss experiences that motivated them to launch a startup.
Balance	14	Founders discuss the alignment of the startup purpose with investors' interest.
Funding	13	Founders discuss their interactions with investors.
Money	13	Founders discuss revenue models and forms of monetization.
Problem identification	13	Founders discuss the problem they address and the need they solve.

Stage 1: Experiences (Personal and Proximate)

In the interviews, founders reflected on their personal experiences during the earliest stage of the genesis of a venture, giving an impression of their passions and feelings. For instance, the founders of Strawberry Energy, Enhancv and mBrainTrain focused on their personal interests and sense of opportunity:

> It all started with a kind of fascination to try to use renewables, like solar energy, in our everyday life.—Strawberry Energy
>
> My co-founder, he had to create a more creative resumé, because he had to find a job in Germany, and he applied for so many companies, and none of them replied. And his resumé was so well written.—Enhancv
>
> We realized that there is no wave brain recorder that is, let's say, tight (...) We also realized that there's no real technological obstacle to its existence.—mBrainTrain

Founders also tried to generalize and relate to broader societal and/or environmental issues. For instance, one of the founders of Grajdanite

Table 5.3 Timeline of second-order categories leading to business model innovation

Timing	Category	Second-order category	Meaning
1	Experience Problem identification	Problem identification	Founders report personal experiences, identify opportunities and find motivation to solve the problems they encounter.
2	Purpose	Purpose identification	Founders connect the problem and their personal identity, making the case for a social/environmental purpose.
3	Teamwork Infrastructure	Team formation	Founders built a team that shares their purpose and complements their skills, usually from their close circle of relatives and friends.
4	Legitimation Money	Interaction with market stakeholders	Founders look for the approval of customers and experiment with monetization forms.
5	Funding		Founders look for the approval of investors and strive to show some early signs of success that is relevant for investors.
6	Balance	Business model innovation	Founders look for a balance between the social/environmental purpose of their endeavours and the interests of market stakeholders that give rise to their business model.

witnessed poor driving and identified this as an educational problem. One of the founders of Playground Energy reflected on his experience as a parent with an increasing concern for children's health as a result of reduced physical activity in childhood. The founder of Farmhopping realized that she was unaware of the quality of the food she bought. At this stage, the startup founder rationalized a situation and honed in on a problem they had encountered that could be experienced by other people in the same situation. Accordingly, founders not only reflected on their own experiences in respect of their enthusiasms and concerns, but also thought about how these personal experiences affected others in their social networks. Frequently, issues were directly extrapolated to family members, while in other cases, the reference group consisted of those in the proximate social context, such as friends or acquaintances:

> Our kids are getting less physically active and more into computers, which is not bad. (…) I mean I have a kid, so they're really very grateful for computers and mobile phones and so on (…) but they need to have a balance.—Playground Energy
>
> We usually go for problems that you see as customers; me, our clients and co-founders, we were uninformed about the food that we bring home and its origin.—Farmhopping
>
> One day, one of the guys [friends] came (…) and he said: 'Guys I had an accident on the road. There was a guy who cut me up. He told us he [the guy from the accident] was alone in the lane and driving a very expensive car, and he seemed to be a very decent person, but somehow he did something annoying to him and, let's say, to ten other people on the road. He [the friend] said he [the guy] knocked on the window, when they stopped next to each other at a traffic light, and he said the guy looked so disturbed that he [the friend] locked the door and closed the windows.—Grajdanite
>
> The problem we're trying to address is the fact that there isn't enough space to write, enough places to write in the office or in school or at home. And also, the idea that you don't feel free to write on the wall.—Escreo

These testimonials are consistent with earlier works on startup creation that emphasize entrepreneurs' motivations for new venture creation, which are usually attributed to the individual's background, experiences and attitudes (Gartner, 1985; Gibb & Ritchie, 1982). The venture creation intent is conceived as a state of mind that describes the thought processes of the entrepreneurs; it precedes action and helps entrepreneurs-to-be to envision the establishment of a new business (Bird, 1988; Learned, 1992).

Our results suggest that the events and experiences that shape the passions and personal identities of the founders are inseparable from the motivations and intentions of these entrepreneurs to create a new venture. We concur that the initial stage of business creation is evolutionary and not instantaneous and acknowledge the influence of personal characteristics but also significant contextual factors in venture creation intent. The role of the proximate social context of the entrepreneur in venture creation intent is confirmed (Forbes, 1999; Shane & Venkataraman, 2000) and exposes the evident limitations of research setups that focus solely on the role of the entrepreneur (Shook, Priem, & McGee, 2003), obliviating experiences lived by family members, friends and colleagues.

Stage 2: Purpose

Having recognized a problem or an opportunity based on personal or proximate experience, and having acknowledged it as something relevant to other social groups, entrepreneurs identified themselves as people capable of taking action. At this stage, entrepreneurs-to-be evaluated business ideas with the aim of creating solutions for the societal or environmental issues previously identified, ending up with the creation of a cognitive schema of the initial business model, based on their understanding of purpose. These solutions represented value propositions permeated with the identities of the founders. At this stage the founders' identities tended to shape the value proposition along a continuum ranging from personal economic profit to a stronger societal purpose related to different impact domains.

In the case of Grajdanite and Strawberry Energy, the societal purpose and the social value added related to the problems of city living.

> We can actually make people think about how they drive if we flag it up to them, and we decided to do this in a very positive way; solving this kind of problem has to be managed in a very civilized way. So, all of us united around this idea, pretty much every member gave suggestions for features that were added to the platform 'Citizens' [Grajdanite].—Grajdanite
>
> The main mission is to accelerate smart city development. Our main focus is on outdoor places, so we are developing this to improve outdoor places in smart cities.—Strawberry Energy

In the case of Farmhopping, Escreo, Playground Energy and Enhancv, purpose was defined in relation to a specific audience: for Playground Energy, the audience was children; Escreo targeted creative and learning communities; Farmhopping focused on small farmers and urban families; and Enhancv served job seekers:

> What we do is motivate kids to play more in the playground. The way we do it is by using some of the kids' energy in the playground, turn it into electricity, and then we use this electricity purely to entertain them and to motivate them to play more.—Playground Energy
>
> What we are trying to do is to present additional functionalities for the space around us, to optimize the space on the one hand and then on the other to help people be more creative and productive during work and at home or in the school during the teaching process and the educational process. So, we believe that our solution is innovative, and it helps people feel more or less like children.—Escreo
>
> [The idea is] to enable farmers to reach the market more directly and to allow customers in the city to have access to food grown locally by farmers in their local area. (...). We started from the idea of supporting farmers and we built this business model around it to make it possible. (...). We wanted to find a way to use technology to bring farmers and consumers closer together.—Farmhopping
>
> It's much easier for you to sell it (with a purpose) to anyone, because it becomes a cause. Secondly, it makes your business more meaningful to yourself and to people who are working with you, because you have the feeling that you're changing something. You see the whole picture. It's not just that you help a guy to find a job at Spotify, but you completely change his life, literally.—Enhancv

On the other hand, mBrainTrain founders leveraged their knowledge and passion for innovation to take advantage of a market opportunity: the lack of compact brainwave recorders:

> And that's why we created it, because we knew the market, we knew the needs, because we were researchers, and so, I guess we anticipated it pretty well.—mBrainTrain

In sum, founders of startups that have a social/environmental purpose identify with that purpose and/or take advantage of the associated market opportunity. Their sensemaking can be perceived as having a more or less personal economic versus societal focus. At both extremes, Grajdanite had a pronounced societal purpose, whereas mBrainTrain started with a more pronounced personal economic focus.

Our findings are consistent with the forging identity process uncovered by Gioia, Price, Hamilton and Thomas (2010). They argue that entrepreneurs play a critical role in the formulation and articulation of the guiding vision of a startup and that the founders' vision delimits the boundaries within which the organizational identity is formed. This idea is consistent with our results: based on personal and proximate experiences, founders form an intention to create a business that is delineated by the founders' purpose. Arguably, this purpose represents what is central, enduring and distinguishable and serves as a reference for members when they are acting on behalf of their organization. The evidence also highlights the need for an understanding of the effects of founders' identities on different aspects of an emerging venture, including team formation, interaction with market stakeholders and business model innovation.

Stage 3: Team Formation

Team formation is a process that hinges on both shared identity and skill complementarity. In the case of Grajdanite and Strawberry Energy, the core team was built on a strong shared identity:

> We gathered 10 people who were, let's say, like-minded and were willing to do something interesting.—Grajdanite
>
> So, all of us united around this idea. We changed a lot of things, pretty much every member, every person contributed suggestions.—Grajdanite
>
> … [a member] gave this initial idea, because he has an IT background in product development. He still is a business development guy for a software company. … [another member] also had an IT background. I didn't have an IT background, but I have a background in business and marketing, and the other guys had the same profile as me.—Grajdanite
>
> And the main inspiration came, so to say, from our passion for trying to find a clever way to use clean energy to try to promote sustainability to engage more people to use solar energy.—Strawberry Energy

In the case of Farmhopping, there was an intentional balance between shared identity and founders' skills:

A very important aspect is why we should be the people to solve this problem, what our expertise and our motivations are, why we are the ones to solve it.—Farmhopping

Team formation is a complex balancing act, as it also involves the need for critical skills that must be acquired using the resources the startup has to hand. Startups don't normally possess abundant resources, and most of these resources are spent on attracting talent. Therefore, the team formation stage is pivotal for the subsequent development stages of the venture. In the case of Grajdanite, the promise of social impact motivated the (remaining) founders to work for free for an extended period of time, delivering societal value. A co-founder in Playground Energy was said to not receive less than the opportunity cost of his time.

After we decided to develop a product and everything around it, not everyone could manage to devote time to it because this was something extra which we did in our spare time, so during this time we went from seven people to just five and then four and then just three. It was me [and two more members], and the rest of the guys, you know, they would help us from time to time but not for the operation of the business, like every day.—Grajdanite

Social networking definitely helps. I mean, there are specific skills or capacities that you feel your company needs, in addition to the funding, to continue growing. (…). We are looking, we don't have enough skills in electrical engineering and programming (…) We need skills there [analytics]. He [a co-founder] is not up for the money.—Playground Energy

Most early-stage ventures, however, competed in the market for talent, and new team members were recruited on competitive terms:

It is very hard for us to find people that are willing to put in not just eight hours of their day, but we look for people that are willing to do this as, you know, their life's mission. I'm not willing to employ anybody that does not perceive his job as his life mission of the moment. So, it's very hard for us to find the appropriate talent.—Melissa Climate

With the investment, we plan to expand the team. We have already approached and have already recruited the fifth member of our team.—Escreo

The key expertise that we need to grow is digital marketing, this is how we reach our customers; it's mostly through digital marketing.—Farmhopping

We found that a strong identification with identity was present in the teams of new ventures as advanced by Fiol.[93] Founders selected new team members, experienced some of them leaving, while others stayed despite not receiving payment for their input, thus shaping the organizational identity of the new venture. Arguably, internal cohesion builds on team members who share the same organizational identity and act accordingly; making sense of their behaviours can lead them to develop a strong sense of purpose. For instance, more than 50% of the founders of Grajdanite, a strong societal purpose-driven startup, left the venture because they were unable to forego their private economic interest. Nevertheless, sharing the same purpose resulted in a level of cohesion that helped the remaining founders maintain the connections with those who left, occasionally benefitting from their talent and time (a free resource).

These observed behaviours are consistent with the process of construction of long-term identity discussed by Gioia et al. (2000). Typically, team members compare the opportunities presented by early-stage ventures with their opportunity cost, taking into account the importance of organizational purpose when making this comparison. Alignment between the personal identity of the team members and the promise of the societal/environmental purpose can make the team more cohesive and less susceptible to changes in the external labour opportunities of team members.

Stage 4: Interaction with Market Stakeholders

At this stage, the startup team's identity interacted with the interests of key market stakeholders, such as customers and investors. The process of market validation of the startup's value proposition is referred to as 'product–market fit'. Potential investors evaluated the promise that the early-stage venture held in terms of the expected returns on their investments. Startup founders now experienced how potential investors prioritized aspects that they themselves had not considered as essential in the process of personal and organizational identity building. Similarly, some of the founders' hypotheses related to customer preferences and behaviours were overturned at this time.

At this stage, we observed that early-stage ventures interacted with early adopters and investors to test whether the problem was well addressed, whether the solution matched the stakeholders' interests and whether the social or environmental purpose contributed additional value and differentiation. In essence, the organizational identity of the early-stage venture, as a function of the purpose addressed, interacted with stakeholders' interests to gain validation. This process usually yielded essential feedback that founders used to update the startup identity and the associated business model that was crystallized in the subsequent stage.

Interaction with Customers

In the case of Grajdanite, the initial interaction with their customers was conditioned by an assumption regarding the preferences and behaviour of the latter. The startup founders believed that customers would not accept a profit-driven model in a startup that relied on altruism and social collaboration. This assumption turned to be wrong.

> So about one year ago we had concerns that if we have customers such a municipality, this will create a very bad image for the organization; but actually, the opposite happened. Actually people, you know, feel kind of they were expecting this to happen, and we don't worry anymore about it.—Grajdanite
>
> It is the highest priority [the social purpose]. We saw many startups trying to do what we are doing and they failed because they were focused on the money in the first place, (...) I remember people telling us: 'you know what you guys do, you can get money from traffic police and from local authorities because you are making a huge optimization, you're saving them the cost of paper, and costs of delivery', etc., but we were not looking for these [opportunities] at that moment.—Grajdanite

The learning about how to think about monetization came with interactions with stakeholders. Entrepreneurs interacted with many different classes of customers to extract the information they needed and later made sense of it, shaping the identity of this early-stage venture:

> We tried working with companies to monitor their fleet. After that we tried to offer people [to pay for] some extra services in a purchase, but it didn't work out.

> Then we went to insurance companies to use their data on traffic violators and manage the risks of the drivers they insure, but it didn't work out either.—Grajdanite

On the other hand, from their interactions with stakeholders, Farmhopping and Playground Energy learnt how to properly communicate with and relate to their customers, and which segments were the right target groups. These interactions with customers, however, did not have a defining impact on the startup organizational identity:

> Another barrier is that we want people to buy food online, which in itself is something that's still not very popular, not just in Bulgaria, but this is something that has been a problem [in most places]. Many online businesses have grown but the online food market has stagnated, has been trying to find its feet, and this has been a problem.—Farmhopping
>
> … one of the main problems is finding the right type of farmers, (…) they should be able to provide enough, they shouldn't be just providing locally or for their neighbours. They should be able to go to a larger market.—Farmhopping
>
> [Our] first buyers were people we knew from before, and scientists, they knew what we were working on, and they were kind of anticipating or expecting this device. (…) However, this turned out to be a very hard aspect, and it turned out that the scientific community was also not in complete agreement, 100% percent, in the procedures that should be applied to that visual neuro kit.—mBrainTrain
>
> This is quite a complex marketing business that we have. We are pushing a lot into B2B. On the other hand, in order to build our brand we also work with some partners like utility companies and have our own playgrounds which are with our brand.—Playground Energy

Interactions with Investors

Generally, investors were equally as important or more important for the evolution of startup founders' identity than customers. All founders realized that investors were primarily motivated by the potential returns on their investment and realized the importance of an alignment of investors' interests and the startup's organizational identity. Startup founders had different perceptions about the balance between private economic profit and societal purpose in the decision process of investors, but they never doubted that return and profit were leading aspects.

> Historically, [investors] usually look for other types of industries. So we didn't have so many [investors] who focus on our area but that's changing. More and more people are coming to invest in smart cities, internet of things, in infrastructure companies, so that's good.—Strawberry Energy
>
> The job of an investor is not to care about [the] society. The job of an investor is to give money.—mBrainTrain
>
> I don't think that investors are really paying attention to these kinds of problems [social and environmental] or opportunities. They would be more focused on the product, or on how to help you find the right people to help you with building the right product or the business.—Escreo
>
> I wouldn't say that it is very important for them [investors], what our company culture is like as long as we deliver the results I promised them. Our newest stakeholder is a very big fan of our culture, he is really interested... and he is giving me a lot of advice.—Enhancv
>
> Well, they [investors] were very interested in my story I would say, because I don't think they had seen something like this back then. But at that time, we didn't have any customers and I think they didn't see any business behind it, so they didn't support us and they were very clear as to why.—Grajdanite
>
> We met other investors, and the very same thing happened. They didn't see a business model there; there was no traction. They wanted to see the hardside. However, on the social part, all the people that we met were very positive. They would say 'This is very good thinking, we as people, as a society, need something like this'.—Grajdanite

Stage 5: Balancing Purpose and Return in Business Model Innovation

The balancing of investors' interests and the identities of startups resulted in a process of continuous adaptation of the business model of early-stage ventures. The aspects that were most affected by this process included the way startups monetized, the way they delivered value and the way they built relationships with customers. At this stage, founders experienced a need for alignment between the interests of market stakeholders and their societal or environmental purpose which was reflected in their organizational identity.

Along the continuum defined by personal economic interest and societal purpose, market stakeholders and startup founders went through a process of convergence. In this process, startup founders learnt about the interests of market stakeholders and adapted the organizational identity of their ventures, revising their cognitive schema and the associated business model. The literature on cultural change proposes two mechanisms for this process: *substitution* and *addition*. For instance, if a startup's identity is strongly driven by societal purpose, founders can substitute existing or add new behaviours that take into account the private economic interest of some stakeholders, changing the identity of the venture and de facto creating a new business model by altering the cognitive schema of the startup team.

For instance, in the case of Farmhopping, the founders realized that the purchasing decisions of households were determined more by convenience and tradition and less by the organic attributes and the locally grown nature of food. This changed the way Farmhopping communicated its value added, emphasizing the convenience of buying from local farmers over the internet instead of going to the supermarket. Additionally, startup founders focused on approaching investors who shared their environmental mission:

> It's [important] to convince the customer and to show them the advantage of buying directly from the farmers instead of going to the supermarket. We are looking for investors who are interested in this industry that we are working in, in the food industry and in the digital markets, and we're currently fundraising from investors who have already invested in Farmhopping.—Farmhopping
>
> Now [social or environmentally-driven businesses] are getting more visibility, and I think with this, more investors will be available.—Farmhopping

Strawberry Energy, Playground Energy and Escreo learnt to approach market segments that were particularly sensitive to their environmental and societal missions to make a better match with investors and customers and to develop a more successful business model. Strawberry Energy, for example, began with the founders' fascination with renewable energy as a personal passion that latterly connected to the startup purpose of environmental protection. Interactions with potential buyers shaped their sensemaking process, turning personal passion into environmental purpose and validating the need to commit to this purpose as a valuable element of differentiation.

> It would have been probably easier, and even in some cases would have had more economic benefits, for example, for some of the lighting devices to be powered by the regular grid. They would be able to offer more features and basically we would sell more of them. But we didn't pursue that avenue because we wanted to be 100% clean.—Strawberry Energy

Playground Energy and Escreo went through a similar adaptation of their founders' cognitive schema. The founders discovered that their personal passions related to a societal purpose and built their organizational identity around it. Managing the balance between this societal purpose and the return-driven interests of the investors helped these startups come up with a business model that served as a reference point for making well-reasoned decisions as well as craft expectations about the future:

> We built our brand to build awareness and to prove our concept, and to really motivate more of our partners and producers of playground equipment to really keep working on [this] and pushing. This is a very new product, nobody in the world is competing with us, it's really a new concept and it takes time [to establish it on the market].—Playground Energy

Moreover, Escreo has been continuously learning about the value they deliver to customers, adapting their cognitive schema as new opportunities seem to emerge in their interactions with clients:

> And we are trying now to open [the product] to schools as well as other educational institutions. (…) We have some client schools. For example, the teacher was examining a student. And the student was asked to stand up next to the Escreo walls to write something there. The teacher actually told us that the student wasn't as anxious when he was writing on the side wall as when he was writing on the front wall.—Escreo

A similar adaptive process occurred in the case of Melissa Climate:

> So, our value proposition really is comfort, everything else we offer them (customers) is kind of an [additional] benefit. But for me it's not just an [additional]

> benefit, it [environmental preservation] is one of the main drivers of why I am doing this.—Melissa Climate

These examples illustrate the way entrepreneurs identify with a social or an environmental purpose and turn it into an additional element of the value they deliver in a process shaped by market stakeholders. Entrepreneurs modified their sensemaking process in their interactions with market stakeholders and that gave rise to startup identities that were not only aligned with the personal identities of founders, but also accounted for the interests of market stakeholders. This process gave startup founders an opportunity to make sense of their evolving business model in relation to aspects such as market segments targeted, value proposition and selection of investors.

Occasionally, the balance between startup identity and stakeholder interests was very explicit in the sensemaking process experienced by entrepreneurs:

> I personally consider all [social purpose and private economic interests] to be needed …, then the question comes of whether you are making your decision using business [logic] and if this startup is ever going to start making money. Like, even if you have a great social impact. What if you don't start making money? And then, what's special socially, it's not there anymore. And from the investor perspective, you lost money.—mBrainTrain
>
> What works now is services supplied to local authorities and obviously we have the biggest value there. So, right now we are focused on, let's say, starting to work with more and more local authorities because this gives us [financial] stability to continue doing what we're doing. We want to move into other problem areas in our society such as violence, corruption, you name it.—Grajdanite
>
> There was an angel investor who offered us a grant, you know, we don't talk much about how much exactly, but it was offered as a grant, and we had several meetings with him. But somehow our visions didn't match, so we didn't move any further and we were pretty disappointed … but again, if the visions of both sides don't match it's not worth it in my opinion.—Grajdanite

Taking the example of Grajdanite, the startup that had the strongest inclination towards societal purpose, the founders realized that their value added did not need to be seen as a form of altruism or a free service. After interacting with market stakeholders, the founders understood their value proposition as the optimization of the administrative work of local authorities. This realization changed the sensemaking of the founders, and they started to charge for their services, which subsequently enhanced their economic value and shaped their business model to create a different balance between the interests of users, customers, investors and their own purpose. The revision of their startup identity enabled the founders to see an opportunity to create organic growth by serving a new customer, local authorities. After the startup gained traction, investors could see Grajdanite had a business model that offered an investment opportunity whilst being aligned with the social mission that motivated the creation of the venture in the first place.

Organizational members' deep identification with what they value and with the outcomes they wish to produce brings coherence to any organization including early-stage ventures that pursue a social mission. Therefore, the genesis of early-stage ventures cannot be understood properly unless the founders' sensemaking of what they do is combined with the interests of market stakeholders, as these represent a reflection of a larger set of contextual norms.

Conclusion

In the course of this research, we have discovered that the process of business model innovation unfolds as a natural step in the genesis of startups with social and environmental purpose and that founders' identities together with market stakeholder interests shape this balance. The transforming context of the Balkans provided us with a unique opportunity to explore the sensemaking processes of startup founders' when it comes to early-stage ventures with societal purpose. In the environment of the historic swing after 1989 towards more individualistic societal institutions, we found that founders of early-stage startups were able to identify opportunities for value creation when bringing societal purpose to the fore. This new slant towards societal orientation among entrepreneurs is indicative of their awareness that some level of collectivism is necessary and beneficial. In addition, market stakeholders such as clients and investors validated this premise during the process of business model innovation.

This chapter examined how the business models of startups with a social or environmental purpose came into existence in the context of a transforming economy. Unlike previous research that used preexisting definitions and frameworks to define enterprises with societal purpose, our study relied on investors' judgements in selecting the startups of interest. In doing so, we explored the genesis of early-stage ventures and uncovered business model innovation as a process that balanced the founders' identity and sensemaking with market stakeholder interests. We found that discussions about the relative importance of these two drivers of business model innovation are futile and unnecessary because business models emerge as a product of both, irrespective of the predominant initial rationalization. Using a phenomenological analysis of seven early-stage ventures that were identified by their investors as possessing a clear social or environmental mission, we identified the communalities that characterized the process of business model creation.

Our research strategy allowed us to identify five stages in the genesis of early-stage ventures with societal purpose, the associated actions and key constituencies. Additionally, the interpretation of the evidence allowed us to argue that many of the discussions in social entrepreneuring that try to prioritize between founders' identity and purpose and investors' interests are ill-grounded as these are characterized by chicken-and-egg types of interdependencies.

Although there exists a disconnect between the view of social entrepreneuring as driven by societal purpose and investor rationale, our evidence shows that this perceived trade-off is false, at least in the case of early-stage ventures with a societal mission in a transforming economy. Viable business models emerge out of the combination of the two rationalizations rather than from the choice of one over the other. Our findings, therefore, empirically corroborate the conjecture that the initial priority of one rationalization over the other is irrelevant because viable business models emerge out of the balance of the two, irrespective of the starting point.

Our findings also revealed that although the dominant starting rationalization is irrelevant for the process of business model innovation, it is more common for founders to augment the societal mission element of their startup identity once there is a convincing return-driven element rather than the other way around. In fact, it is possible to conjecture that startups that are strongly mission-driven as a consequence of the founders' identities require more time to experiment and figure out how to incorporate investors' interests in their business models and, as a consequence, their founders need to be more patient. Subsequently, the personal identities of the founders and the startup identity of the mission-driven venture evolve and adapt, by adding and/or substituting behaviours. Our results also point to a possible bias in the literature of social entrepreneurship that is based on in-depth case studies but does not include the uncovering of founders' identities and the associated process of sensemaking. Unless studies of social entrepreneurship include an investigation of entrepreneurs' identity and sensemaking and allow for a sufficiently long timeframe, the societal element in early-stage ventures will be systematically underreported.

This research offers several contributions to the literature on entrepreneuring. First, it adds to the growing literature on business model innovation by combining two disconnected views on entrepreneuring—social entrepreneurship and investor rationalization. In doing this, we part from the more traditional approach to studying business models that focuses on specific resources (Meyskens, Robb-Post, Stamp, Carsrud, & Reynolds, 2010; Seelos & Mair, 2005; Seelos, 2014), and which emphasize the role of personal and organizational identities and sensemaking. Unlike previous studies, our approach focuses on the role of interactions with stakeholders and the balancing of different rationalizations as instrumental for the emergence of a viable business model. We explicitly allow for a different balance between societal purpose and private economic interest rather than treating the societal component as an 'all or nothing' feature of the business model.

This chapter draws on research in organizational sociology and cultural change as advanced by Albert and Whetten (1985) and Fiol (2001) to argue that startup founder identities can be understood as the outcomes of a process of sensemaking of founders' actions on the basis of an intermediate set of beliefs. From this perspective, the genesis of early-stage ventures cannot be understood unless the founders' sensemaking of what they do is taken into account together with the interests of market stakeholders, as these represent a reflection of a larger set of contextual norms. Clearly, the business model innovation process of early-stage ventures is impossible to describe if either founders' identities or market stakeholders' interests are ignored.

Finally, our research approach complements the growing number of literature strands that call for entrepreneuring to be studied as a social phenomenon (Aldrich & Zimmer, 1986) and that conceive the process of founding a business venture as rooted in social interactions. In fact, these different strands of literature have led to the emergence of complementary research themes, such as the influence of social networks (de Carolis, Litzky, & Eddleston, 2009); social capital (Davidsson & Honig, 2003; de Carolis & Saparito, 2006; Gedajlovic, Honig, Moore, Payne, & Wright, 2013); employers (Sørensen & Fassiotto, 2011); mentors (Eesley & Wang, 2017); peers (Nanda & Sørensen, 2010; Qin & Estrin, 2015) and role models (Lafuente, Vaillant, & Rialp, 2007; Roach & Sauermann, 2015; Wyrwich, Stuetzer, & Sternberg, 2016). Our definition of business model innovation as a balancing act between founders' identities and market stakeholder interests is complementary to all of these perspectives.

Our goal in this chapter was to uncover how the business models of early-stage ventures with societal purpose come into existence and how in this process market stakeholders such as return-maximizing investors and utility-maximizing clients shape the product–market fit of these startups and, consequently, the balance between private economic and societal purpose. The five stages we identified in this process provide important theoretical and empirical insights and at an organizational level allow us to see business process innovation as a balancing act between organizational resources (founders' identities) and market conditions (investors' and customers' interests). Systematically studying the differences across these market conditions is a valuable avenue for research.

In addition, disentangling the specific mechanisms through which interacting stakeholder interests shape startup identity and consequently its business model remains to be fully understood, and the complexity of the task is significant. One possible venue for advancing this stream of research was proposed by McCloskey (2006) who vividly illustrated how bourgeois virtues gave rise to entrepreneurial initiatives that were indispensable to the economic takeoff in 1800s Europe. In this historical period, entrepreneurship and innovation blossomed and triggered an unprecedented increase in the economic wellbeing of societies because, according to McCloskey, of the particular balance among entrepreneurial ('bourgeois') virtues that originated in three distinct spheres of living: economic self-interest, social regulation and the theological dimension. By showing the interaction between entrepreneurs and different market stakeholders, this research offers a framework for analysing entrepreneurial predispositions in line with that of McCloskey (2006). From this perspective, the distinction that we make between personal economic interests and societal purpose does not appear to provide sufficient guidance as to how to go beyond the understanding that 'individuals need the nurture of groups that carry a moral tradition reinforcing their own aspirations' (Bellah, Madsen, Sullivan, & Swidler, 2007). The articulation of a more powerful and actionable framework that encompasses both the entrepreneur's identity and sensemaking and their social and theological contexts thus remains a future endeavour for social scientists.

References

Abdelkafi, N., & Täuscher, K. (2016). Business models for sustainability from a system dynamics perspective. *Organization & Environment, 29*(1), 74–96. https://doi.org/10.1177/1086026615592930

Acs, Z., & Storey, D. (2004). Introduction: Entrepreneurship and economic development. *Regional Studies, 38*(8), 871–877. https://doi.org/10.1080/0034340042000280901

Aidis, R., Welter, F., Smallbone, D., & Isakova, N. (2007). Female entrepreneurship in transition economies: The case of Lithuania and Ukraine. *Feminist Economics, 13*(2), 157–183. https://doi.org/10.1080/13545700601184831

Albert, S., & Whetten, D. (1985). Organizational identity. *Research in Organizational Behavior, 7*, 263–295.

Aldrich, H. E., & Zimmer, C. (1986). Entrepreneurship through social networks. In C. Zimmer (Ed.), *The art and science of entrepreneurship* (pp. 3–23). Cambridge, MA: Ballinger.

Alvord, S., Brown, L., & Letts, C. (2004). Social entrepreneurship and societal transformation. *The Journal of Applied Behavioral Science, 40*(3), 260–282. https://doi.org/10.1177/0021886304266847

Ashforth, B., & Mael, F. (1989). Social identity theory and the organization. *The Academy of Management Review, 14*(1), 20. https://doi.org/10.2307/258189

Aspara, J., Lamberg, J., Laukia, A., & Tikkanen, H. (2013). Corporate business model transformation and inter-organizational cognition: The case of Nokia. *Long Range Planning, 46*(6), 459–474. https://doi.org/10.1016/j.lrp.2011.06.001

Aulet, W., & Murray, F. (2013). A tale of two entrepreneurs: Understanding differences in the types of entrepreneurship in the economy. *SSRN Electronic Journal.* https://doi.org/10.2139/ssrn.2259740

Barney, J., Bunderson, J., Foreman, P., Gustafson, L., Huff, A., Martins, L., et al. (1998). A strategy conversation on the topic of organization identity. In A. Whetten & P. Godfrey (Eds.), *Identity in organizations: Building theory through conversations* (pp. 99–168). Thousand Oaks, CA: Sage.

Battilana, J., & Dorado, S. (2010). Building sustainable hybrid organizations: The case of commercial microfinance organizations. *Academy of Management Journal, 53*(6), 1419–1440. https://doi.org/10.5465/amj.2010.57318391

Battilana, J., Lee, M., Walker, J., & Dorsey, C. (2012). In search of the hybrid ideal. *Stanford Social Innovation Review, 10*(3), 50–55.

Baum, J., Calabrese, T., & Silverman, B. (2000). Don't go it alone: Alliance network composition and startups' performance in Canadian biotechnology. *Strategic Management Journal, 21*(3), 267–294. https://doi.org/10.1002/(sici)1097-0266(200003)21:3<267::aid-smj89>3.0.co;2-8

Berends, H., Smits, A., Reymen, I., & Podoynitsyna, K. (2016). Learning while (re) configuring: Business model innovation processes in established firms. *Strategic Organization, 14*(3), 181–219. https://doi.org/10.1177/1476127016632758

Bellah, R., Madsen, R., Sullivan, W., & Swidler, A. (2007). *Habits of the heart: Individualism and commitment in American life.* University of California Press.

Bird, B. (1988). Implementing entrepreneurial ideas: The case for intention. *Academy of Management Review, 13*(3), 442–453. https://doi.org/10.5465/amr.1988.4306970

Birkinshaw, J., & Goddard, J. (2009). What is your management model?. MIT sloan management review. *Cambridge, 50*(2), 81–90.

Bocken, N., Rana, P., & Short, S. (2015). Value mapping for sustainable business thinking. *Journal of Industrial And Production Engineering, 32*(1), 67–81. https://doi.org/10.1080/21681015.2014.1000399

Boons, F., & Lüdeke-Freund, F. (2013). Business models for sustainable innovation: State-of-the-art and steps towards a research agenda. *Journal of Cleaner Production, 45*, 9–19. https://doi.org/10.1016/j.jclepro.2012.07.007

Bosma, N., & Harding, R. (2007). *Global Entrepreneurship Monitor: GEM 2006 results*. GEM.

Casadesus-Masanell, R., & Ricart, J. (2010). From strategy to business models and onto tactics. *Long Range Planning, 43*(2–3), 195–215. https://doi.org/10.1016/j.lrp.2010.01.004

Casadesus-Masanell, R., & Zhu, F. (2010). Strategies to fight ad-sponsored rivals. *SSRN Electronic Journal*. https://doi.org/10.2139/ssrn.1476530

Casson, M., & Wadeson, N. (2007). The discovery of opportunities: Extending the economic theory of the entrepreneur. *Small Business Economics, 28*(4), 285–300. https://doi.org/10.1007/s11187-006-9037-7

Chesbrough, H. (2010). Business model innovation: Opportunities and barriers. *Long Range Planning, 43*(2–3), 354–363. https://doi.org/10.1016/j.lrp.2009.07.010

Chesbrough, H., & Rosenbloom, R. (2002). The role of the business model in capturing value from innovation: Evidence from Xerox corporation's technology spin-off companies. *Industrial and Corporate Change, 11*(3), 529–555. https://doi.org/10.1093/icc/11.3.529

Cope, J. (2005). Researching entrepreneurship through phenomenological inquiry: Philosophical and methodological issues. *International Small Business Journal, 23*(2), 163–189. https://doi.org/10.1177/0266242605050511

Dahan, N., Doh, J., Oetzel, J., & Yaziji, M. (2010). Corporate-NGO collaboration: Co-creating new business models for developing markets. *Long Range Planning, 43*(2–3), 326–342. https://doi.org/10.1016/j.lrp.2009.11.003

Davidsson, P. (1991). Continued entrepreneurship: Ability, need, and opportunity as determinants of small firm growth. *Journal of Business Venturing, 6*(6), 405–429. https://doi.org/10.1016/0883-9026(91)90028-c

Davidsson, P. (1995). Culture, structure and regional levels of entrepreneurship. *Entrepreneurship & Regional Development, 7*(1), 41–62. https://doi.org/10.1080/08985629500000003

Davidsson, P., & Honig, B. (2003). The role of social and human capital among nascent entrepreneurs. *Journal of Business Venturing, 18*(3), 301–331. https://doi.org/10.1016/S0883-9026(02)00097-6

Davidsson, P., & Wiklund, J. (1997). Values, beliefs and regional variations in new firm formation rates. *Journal of Economic Psychology, 18*(2–3), 179–199. https://doi.org/10.1016/s0167-4870(97)00004-4

de Carolis, D. M., Litzky, B. E., & Eddleston, K. A. (2009, March). Why networks enhance the progress of new venture creation: The influence of social capital and cognition. *Entrepreneurship Theory and Practice, 33*, 527–545.

de Carolis, D. M., & Saparito, P. (2006). Social capital, cognition, and entrepreneurial opportunities: A theoretical framework. *Entrepreneurship Theory and Practice, 30*(1), 41–56.

Doz, Y., & Kosonen, M. (2010). Embedding strategic agility: A leadership agenda for accelerating business model renewal. *Long Range Planning, 43*(2–3), 370–382. https://doi.org/10.1016/j.lrp.2009.07.006

Dutton, J., & Dukerich, J. (1989). Keeping an eye on the mirror: Image and identity in organizational adaptation. *Academy of Management Journal, 34*(3), 517–554. https://doi.org/10.2307/256405

Dutton, J. E., & Jackson, S. (1989). *Does what you see depend on where you sit?: Linking context and strategic issue categories*. Paper presented at the Academy of Management meetings, Washington, DC.

Eesley, C., & Wang, Y. (2017). Social influence in career choice: Evidence from a randomized field experiment on entrepreneurial mentorship. *Research Policy*. https://doi.org/10.1016/j.respol.2017.01.010

Eleven quarterly report. 2017. Made available in private correspondence with the eleven fund managing partners.

Erez, M., & Earley, P. (1993). *Culture, self-identity and work*. New York: Oxford University Press.

Fauchart, E., & Gruber, M. (2011). Darwinians, communitarians, and missionaries: The role of founder identity in entrepreneurship. *Academy of Management Journal, 54*(5), 935–957. https://doi.org/10.5465/amj.2009.0211

Fiol, C. (1991). Managing culture as a competitive resource: An identity-based view of sustainable competitive advantage. *Journal of Management, 17*(1), 191–211. https://doi.org/10.1177/014920639101700112

Fiol, C. (2001). Revisiting an identity-based view of sustainable competitive advantage. *Journal of Management, 27*(6), 691–699. https://doi.org/10.1177/014920630102700606

Fiol, C., & Dunbar, R. (1990). *Identifying the basis for organizational cohesion: The tight side of loose coupling*. Leonard N. Stern School of Business, New York University.

Forbes, D. (1999). Cognitive approaches to new venture creation. *International Journal of Management Reviews, 1*(4), 415–439. https://doi.org/10.1111/1468-2370.00021

Gartner, W. (1985). A conceptual framework for describing the phenomenon of new venture creation. *Academy of Management Review, 10*(4), 696–706. https://doi.org/10.5465/amr.1985.4279094

Gedajlovic, E., Honig, B., Moore, C. B., Payne, G. T., & Wright, M. (2013). Social capital and entrepreneurship: A schema and research agenda. *Entrepreneurship Theory and Practice, 37*(3), 455–478.

GEM. (2017). *GEM 2016/2017 global report* (pp. 1–180). Global Entrepreneurship Monitor. Retrieved from http://www.gemconsortium.org/report/49812

Startup Genome. (2012). Startup ecosystem report.

Gibb, A., & Ritchie, J. (1982). Understanding the process of starting small businesses. *European Small Business Journal, 1*(1), 26–45. https://doi.org/10.1177/026624268200100102

Gioia, D., Price, K., Hamilton, A., & Thomas, J. (2010). Forging an identity: An insider-outsider study of processes involved in the formation of organizational identity. *Administrative Science Quarterly, 55*(1), 1–46. https://doi.org/10.2189/asqu.2010.55.1.1

Gioia, D., Schultz, M., & Corley, K. (2000). Organizational identity, image, and adaptive instability. *The Academy of Management Review, 25*(1), 63. https://doi.org/10.2307/259263

Goffman, E. (1967). *Interaction ritual*. New York: Routledge.

Grabenwarter, U., & Liechtenstein, H. (2011). In search of gamma—An unconventional perspective on impact investing. *SSRN Electronic Journal*. https://doi.org/10.2139/ssrn.2120040

Gries, T., & Naudé, W. (2010). Entrepreneurship and structural economic transformation. *Small Business Economics, 34*(1), 13–29. https://doi.org/10.1007/s11187-009-9192-8

Harding, R., Brooksbank, D., Hart, M., Jones-Evans, D., Levie, J., O'Reilly, J., et al. (2006). *Global Entrepreneurship Monitor United Kingdom 2005*. London: London Business School.

Haugh, H. (2005). A research agenda for social entrepreneurship. *Social Enterprise Journal, 1*(1), 1–12. https://doi.org/10.1108/17508610580000703

Hausmann, R., & Rodrik, D. (2003). Economic development as self-discovery. *Journal of Development Economics, 72*(2), 603–633. https://doi.org/10.1016/s0304-3878(03)00124-x

Hitlin, S. (2003). Values as the core of personal identity: Drawing links between two theories of self. *Social Psychology Quarterly, 66*(2), 118. https://doi.org/10.2307/1519843

Hockerts, K. (2006). Entrepreneurial opportunity in social purpose business ventures in Social Entrepreneurship. In J. Mair, J. Robinson, & K. Hockerts (Eds.), *Handbook of research in social entrepreneurship*. Palgrave Macmillan.

Hofstede, G. (1980). Motivation, leadership, and organization: Do American theories apply abroad? *Organizational Dynamics, 9*(1), 42–63. https://doi.org/10.1016/0090-2616(80)90013-3

House, R., Hanges, P., Javidan, M., Dorfman, P., & Gupta, V. (2004). Culture, leadership, and organizations.

Hycner, R. (1985). Some guidelines for the phenomenological analysis of interview data. *Human Studies, 8*(3), 279–303. https://doi.org/10.1007/bf00142995

Isenberg, D. (2011). *The entrepreneurship ecosystem strategy as a new paradigm for economic policy: Principles for cultivating entrepreneurship*. Presentation, Institute of International and European Affairs.

Itami, H., & Nishino, K. (2010). Killing two birds with one stone. *Long Range Planning, 43*(2–3), 364–369. https://doi.org/10.1016/j.lrp.2009.07.007

Kimberly, J. (1979). Issues in the creation of organizations: Initiation, innovation, and institutionalization. *Academy of Management Journal, 22*(3), 437–457. https://doi.org/10.2307/255737

Kreiser, P., Marino, L., Dickson, P., & Weaver, K. (2010). Cultural influences on entrepreneurial orientation: The impact of national culture on risk taking and proactiveness in SMEs. *Entrepreneurship Theory and Practice, 34*(5), 959–983. https://doi.org/10.1111/j.1540-6520.2010.00396.x

Lafuente, E., Vaillant, Y., & Rialp, J. (2007). Regional differences in the influence of role models: Comparing the entrepreneurial process of rural Catalonia. *Regional Studies, 41*(6), 779–796.

Learned, K. (1992). What happened before the organization? A model of organization formation. *Entrepreneurship Theory and Practice, 17*(1), 39–48. https://doi.org/10.1177/104225879201700105

Magretta, J. (2002, May). Why business models matter. *Harvard Business Review*, 86–92.

Mair, J., Battilana, J., & Cardenas, J. (2012). Organizing for society: A typology of social entrepreneuring models. *Journal of Business Ethics, 111*(3), 353–373. https://doi.org/10.1007/s10551-012-1414-3

Martins, L., Rindova, V., & Greenbaum, B. (2015). Unlocking the hidden value of concepts: A cognitive approach to business model innovation. *Strategic Entrepreneurship Journal, 9*(1), 99–117. https://doi.org/10.1002/sej.1191

Massa, L., Tucci, C., & Afuah, A. (2017). A critical assessment of business model research. *Academy of Management Annals, 11*(1), 73–104. https://doi.org/10.5465/annals.2014.0072

McCracken, G. (1988). *The long interview*. Newbury Park, CA: Sage.

McCloskey, D. N. (2006). *Bourgeois virtue: Ethics for an age of commerce*. Chicago: The University of Chicago Press.

McGrath, R., MacMillan, I., & Scheinberg, S. (1992). Elitists, risk-takers, and rugged individualists? An exploratory analysis of cultural differences between entrepreneurs and non-entrepreneurs. *Journal of Business Venturing, 7*(2), 115–135. https://doi.org/10.1016/0883-9026(92)90008-f

Meyskens, M., Robb-Post, C., Stamp, J. A., Carsrud, A. L., & Reynolds, P. D. (2010). Social ventures from a resource-based perspective: An exploratory study assessing global Ashoka fellows. *Entrepreneurship Theory and Practice, 34*(4), 661–680.

Minniti, M., Bygrave, W., & Autio, E. (2006). *Global Entrepreneurship Monitor: 2005 executive report*. London: London Business School.

Minniti, M., & Naudé, W. (2010). What do we know about the patterns and determinants of female entrepreneurship across countries? *The European Journal of Development Research, 22*(3), 277–293. https://doi.org/10.1057/ejdr.2010.17

Monitor Institute. (2009). *Investing for social & environmental impact* (p. 86). Monitor Institute (Monitor Deloitte).

Morris, M., Schindehutte, M., & Allen, J. (2005). The entrepreneur's business model: Toward a unified perspective. *Journal of Business Research, 58*(6), 726–735. https://doi.org/10.1016/j.jbusres.2003.11.001

Nanda, R., & Sørensen, J. B. (2010). Workplace peers and entrepreneurship. *Management Science, 56*(7), 1116–1126. https://doi.org/10.1287/mnsc.1100.1179

Neumeyer, X., & Corbett, A. (2017). Entrepreneurial ecosystems: Weak metaphor or genuine concept? In *Advances in the study of entrepreneurship, innovation & economic growth* (pp. 35–45). Retrieved from https://doi.org/10.1108/s1048-473620170000027005

Nielsen, C., & Lund, M. (2014). A brief history of the business model concept. *SSRN Electronic Journal*. https://doi.org/10.2139/ssrn.2579439

Porter, M., & Kramer, M. (2011). The big idea: Creating shared value. How to reinvent capitalism—And unleash a wave of innovation and growth. *Harvard Business Review, 89*, 1–2.

Qin, F., & Estrin, S. (2015). Does social influence span time and space? Evidence from Indian returnee entrepreneurs. *Strategic Entrepreneurship Journal, 9*, 226–242. https://doi.org/10.1002/sej

Reim, W., Parida, V., & Örtqvist, D. (2015). Product–Service Systems (PSS) business models and tactics—A systematic literature review. *Journal of Cleaner Production, 97*, 61–75. https://doi.org/10.1016/j.jclepro.2014.07.003

Rindova, V., Barry, D., & Ketchen, D. (2009). Entrepreneuring as emancipation. *Academy of Management Review, 34*(3), 477–491. https://doi.org/10.5465/amr.2009.40632647

Roach, M., & Sauermann, H. (2015). Founder or joiner? The role of preferences and context in shaping different entrepreneurial interests. *Management Science, 61*(9), 2160–2184. https://doi.org/10.1287/mnsc.2014.2100

Schmutzler, J., Andonova, V., & Diaz-Serrano, L. (2018). How context shapes entrepreneurial self-efficacy as a driver of entrepreneurial intentions: A multilevel approach. *Entrepreneurship Theory and Practice*. https://doi.org/10.1177/1042258717753142

Schwartz, S. (1994). Are there universal aspects in the structure and contents of human values? *Journal of Social Issues, 50*(4), 19–45. https://doi.org/10.1111/j.1540-4560.1994.tb01196.x

Schwartz, S. (2012). An overview of the Schwartz theory of basic values. *Online Readings in Psychology and Culture, 2*(1), 11. https://doi.org/10.9707/2307-0919.1116

Seelos, C. (2014). Theorising and strategising with models: Generative models of social enterprises. *International Journal of Entrepreneurial Venturing, 6*(1), 6–21.

Seelos, C., & Mair, J. (2005). Social entrepreneurship: Creating new business models to serve the poor. *Business Horizons, 48*(3), 241–246.

Shane, S. (1992). Why do some societies invent more than others? *Journal of Business Venturing, 7*(1), 29–46 https://doi.org/10.1016/0883-9026(92)90033-n

Shane, S., & Venkataraman, S. (2000). The promise of entrepreneurship as a field of research. *Academy of Management Review, 25*(1), 217–226. https://doi.org/10.5465/amr.2000.2791611

Shook, C., Priem, R., & McGee, J. (2003). Venture creation and the enterprising individual: A review and synthesis. *Journal of Management, 29*(3), 379–399. https://doi.org/10.1016/s0149-2063_03_00016-3

Short, J., Moss, T., & Lumpkin, G. (2009). Research in social entrepreneurship: Past contributions and future opportunities. *Strategic Entrepreneurship Journal, 3*(2), 161–194. https://doi.org/10.1002/sej.69

Smallbone, D., & Welter, F. (2004). *Entrepreneurship in transition economies: Necessity or opportunity driven?* Paper presented at the Babson-Kauffman Entrepreneurship Research Conference, Babson College, Boston, MA, June.

Smith, J., & Eatough, V. (2006). Interpretative phenomenological analysis. In G. Breakwell, S. Hammond, C. Fife-Schan, & J. Smith (Eds.), *Research methods in psychology*. London: Sage Publications.

Smith, J. A., Jaman, M., & Osborn, M. (1999). Doing interpretive phenomenological analysis. In M. Murray & K. Chamberlain (Eds.), *Qualitative health psychology: Theories and methods*. London: Sage.

Smith, W., Binns, A., & Tushman, M. (2010). Complex business models: Managing strategic paradoxes simultaneously. *Long Range Planning, 43*(2–3), 448–461. https://doi.org/10.1016/j.lrp.2009.12.003

Sørensen, J. B., & Fassiotto, M. A. (2011). Organizations as fonts of entrepreneurship. *Organization Science, 22*(5), 1322–1331. https://doi.org/10.1287/orsc.1100.0622

Sosna, M., Trevinyo-Rodríguez, R., & Velamuri, S. (2010). Business model innovation through trial-and-error learning. *Long Range Planning, 43*(2–3), 383–407. https://doi.org/10.1016/j.lrp.2010.02.003

Teece, D. (2010). Business models, business strategy and innovation. *Long Range Planning, 43*(2–3), 172–194. https://doi.org/10.1016/j.lrp.2009.07.003

Thomas, A., & Mueller, S. (2000). A case for comparative entrepreneurship: Assessing the relevance of culture. *Journal of International Business Studies, 31*(2), 287–301. https://doi.org/10.1057/palgrave.jibs.8490906

Timmers, P. (1998). Business models for electronic markets. *Electronic Markets, 8*(2), 3–8. https://doi.org/10.1080/10196789800000016

Wells, P. (2016). Economies of scale versus small is beautiful: A business model approach based on architecture, principles and components in the beer industry. *Organization & Environment, 29*(1), 36–52. https://doi.org/10.1177/1086026615590882

Wennekers, A., Uhlaner, L., & Thurik, A. (2002). Entrepreneurship and its conditions: A macro perspective. *International Journal of Entrepreneurship Education (IJEE), 1*(1), 25–64 Retrieved from http://Hdl.Handle.Net/1765/15876

Whetten, D. (2006). Albert and Whetten revisited: Strengthening the concept of organizational identity. *Journal of Management Inquiry, 15*(3), 219–234. https://doi.org/10.1177/1056492606291200

Whetten, D., & Mackey, A. (2002). A social actor conception of organizational identity and its implications for the study of organizational reputation. *Business & Society, 41*(4), 393–414. https://doi.org/10.1177/0007650302238775

Williams, C. (2008). Beyond necessity-driven versus opportunity-driven entrepreneurship. *The International Journal of Entrepreneurship and Innovation, 9*(3), 157–165. https://doi.org/10.5367/000000008785096647

Wyrwich, M., Stuetzer, M., & Sternberg, R. (2016). Entrepreneurial role models, fear of failure, and institutional approval of entrepreneurship: A tale of two regions. *Small Business Economics, 46*(3), 467–492. https://doi.org/10.1007/s11187-015-9695-4

Zott, C., & Amit, R. (2010). Business model design: An activity system perspective. *Long Range Planning, 43*(2–3), 216–226. https://doi.org/10.1016/j.lrp.2009.07.004

Zott, C., Amit, R., & Massa, L. (2011). The business model: Recent developments and future research. *SSRN Electronic Journal*. https://doi.org/10.2139/ssrn.1674384

Part II

Case Studies

Introduction

In a series of structured interviews with startup founders from the region, we have tried to identify and summarize the features and characteristics of entrepreneurs that are specific to this part of the world and identify any common traits that make startups from this region successful.

Of the 6500 startup projects that were screened by Eleven and LAUNCHub, the most active investors on the Balkans between 2012 and 2015, close to 200 (or a bit less than 3%) received an investment. From these we have selected six teams, as well as two other companies, to interview in an attempt to uncover any systematic similarities or differences among them, to understand how they are alike or different from startups in other parts of the world and to demonstrate how they work within their teams and with other stakeholders, like mentors and investors. In some of the cases, we have also spoken with their investors or advisors to contextualize the founders' perspective, but to also understand how investors in this part of the world consider and evaluate startups. Furthermore, each of these companies represents a specific case study, and analysing these provides a valuable learning experience we want to share with aspiring entrepreneurs, to give them additional perspectives on and understanding of certain aspects of starting and running a startup; to investors, so that they become aware of the early warning signals and strategies that proved successful; and to policy-makers who want to understand what is needed in order to build vibrant entrepreneurial ecosystems in unexpected latitudes.

We have analysed several key elements in these interviews and have added on top a layer of our own experience in working with over 100 startups in the last five years, coming predominantly from countries in the Balkans and Central and Eastern Europe. These elements are (i) the genesis of a successful startup, (ii) team dynamics and interactions, (iii) the role of advisors and investors in the life of a startup, (iv) the role of the environment and (v) the building blocks for successful fundraising and investment readiness. We have spoken with each founder about these topics and have dived deeper into topics especially relevant to those particular startups, but (we believe) vital for startups in the region in general.

There is a lot of knowledge to be extracted from the interviews or learning outcomes that are relevant for particular projects or stages in the life of a startup, beyond the conclusions we have highlighted, so alongside the case summaries, we advise readers to absorb both the founder and investor interviews to see the different points of view of both parties.

Genesis

People's drive and motivation to start a startup can be very different. Because of media bias, the stories we hear most often are about 'rock-star' founders: people who like the spotlight, who are keen to spend time at conferences and startup events and who are happy to talk to the media and write articles themselves. This puts entrepreneurship in a somewhat illusory light, giving the outside observer the impression that it is a lot of fun, hustling, luck and chance, and maybe a bit of work. It also drives a lot of people towards entrepreneurship without having a clear idea of how hard starting and running a company actually is. We have seen many teams that seem to have come up with an idea out of the blue, without having any prior experience in the domain, hardly researching the market or seriously considering what it would take to actually build their projects—teams that create a good pitch deck, practise their presentations and apply for funding. It should not come as a surprise that, when these teams get rejected for funding, very few invest further time in their projects in order to try a second time once they are more advanced. Those teams that did get the ticket would often be the first to throw in the towel.

Of the interviews we held for this book, five of the teams started their businesses out of necessity, by chance, or because they spotted an opportunity in their previous business and decided that it had a better chance as a startup, rather than as part of a corporation. The cases of Telerik, Coherent Labs, Sensika and FITE are very good examples of that—none of them became an entrepreneur because they were inspired by the story of, let's say, Elon Musk or Steve Jobs, or wanted to be the next Bill Gates.

We believe this is a key ingredient of successful startups: you cannot come up with a great idea just by asking 'what will be the next big thing?' Ideas come when you face a problem or when an opportunity presents itself in a sector you know well or in which you have at least some experience. And if the founders' motivation to start a company is to become rich and famous, it will almost never work.

We cannot offer practical advice here on how to find the next big idea, but we can definitely recommend that you avoid pushing yourself to come up with it. It will come your way when the time is right.

Team

You can read a lot about building and motivating teams in early-stage startups, about stock option plans, team building activities and managing work–life balance, and you can try to apply all of these in your startup. Maybe some of these will work, but there were two major themes that emerged from our conversations with the startup founders in this book:

(i) There is no such thing as corporate culture in a startup, there is only the culture of the founders. Usually the first hires of a startup are made by the founders themselves, so they are chosen either from the founders' personal networks, or share the same values and vision as the founders. This sets the tone for the next hires, and the next, so that at the end of the day, the top-tier people in a startup more or less reflect the mentality, mindset and shared values of the founders. It takes a lot of steps before the founder archetypes are blurred enough so that the next member of the team can be radically different from the first. And probably a few steps before that would be the right time for an organization to start investing in formalizing and maintaining its corporate culture—before that, it is natural. And what is more, if the founders try to impose on their company a culture that is different from their own, it is likely not going to work.

(ii) Teams get inspired when they see their leader by their side in the trenches, working long hours and putting his or her full effort into the endeavour. Sometimes we have seen wannabe founders approach us for funding to hire and pay for a full team, while they plan to do something else and just run the strategic part of the business. And this will never work for a startup.

Investors and Advisors

We hear the term 'smart money' quite often when talking about startup investments, and the thought that naturally occurs is that your investors' team is what puts the 'smart' in the smart money. And they do, although not directly. There are four core advantages the founders we spoke to outlined regarding having an investor:

(i) *Confidence*—when your company gets an investment, this gives your team wings, as there are now more smart people that have bought into your story, believe what you are doing may be successful and are ready to invest real money into it.

(ii) *Runway*—now that you have that cheque in your bank account, you have more time to experiment and get to the next stage, whether that's product–market fit, product development or growth mechanics. You now have the time you need to hit the next milestone and either get more funding or become profitable.

(iii) *Network*—this is probably the most important part, and it is what puts the 'smart' into 'smart money'. Once you become part of your investors' portfolio, their network becomes your network, and you can easily get in touch with the right people to help you solve a problem or close a deal. And what is more, even if those people are not in your investors' direct contacts, just having a prominent investor on board is enough to open the doors to your future clients, partners or hires.
(iv) *Discipline*—investors will ask you to report to them regularly on your financials and key metrics, insist you implement strict corporate governance processes and procedures and question any major decision you take for your startup. So, with investors on board, as Svetlozar Georgiev, the founder of Telerik put it, you move to the next level and become a real company.

And yes, sometimes your investors may have practical knowledge on certain matters that are relevant for your startup (other than term sheets, investment terms and financial management), and they can give you practical advice, but this stems not from their role of investors, but from their role as mentors. As George Petrov, the founder of Coherent Labs, put it, your investors can give you theoretical advice—the 'what' you need to do—but what you really need when you are an early-stage startup is the 'how' exactly to do it.

This is where mentors come in, and they usually come out of the combined network of the startup, its investors and their portfolio companies, so the larger and more relevant to your domain this network is, the better your chances of finding the right person to speak to. And that is why vertically focused investors may be more helpful to their startups than generalists, as their network is concentrated in that particular field.

Another thing that came out of our discussions is that the international diaspora of emigrants from your country can be really helpful and is definitely more likely to engage and walk the extra mile for you than some random contact in that market. People from one country that have become successful abroad have a tendency to be more responsive and supportive to startup founders from that country. As Daniel Tomov, the investor from Eleven, put it, it is probably this positive feeling of guilt that they cannot contribute directly to their own country that drives them to be more helpful to other founders there.

Environment

We asked the founders and investors what they think would have been different for each company, if it had launched in Silicon Valley, expecting to hear that they would have been much more successful there, but we were surprised to hear a different story. People believe that Silicon Valley would give them a better pool of investors and advisors and that their project would probably have developed (succeeded or failed) faster, but no one said that it would have made a significant difference in the outcome of the project (e.g. a project would have failed in the Balkans, but would have been successful in California).

Furthermore, there were factors in the environment in our region, or other places around the world, that were mentioned as clear advantages to Silicon Valley, like loyalty, dedication or a more relevant market, to name a few.

Investment Readiness

We have included this topic in the investor interviews, because we have all too often heard startups ask what the process of applying for an investment is, but not nearly as often what they would need to do to be ready for an investment. And in our experience, if you are at the right stage with the right setup, it is much easier to get funded.

The most important difference and factor in the decision-making of investors seems to be the time the team has spent on their project before seeking external funding. This is natural, because teams that have not spent time and effort on a project are assuming virtually no risk on their own, and it is hard to say if they are doing it because they really believe in it or are just giving it a try. And if it is the latter, teams tend to quit early.

Furthermore, if the startup has spent some time working on a problem, they probably covered some of the first milestones, effectively de-risking the project for the investors, which, of course, would mean a higher valuation, but nonetheless, investors would be much more keen to look at that project than at a team with a PowerPoint presentation but little substance or experience.

Secondly, team history and dynamics seem to be just as important. Although investors don't like single founder teams, they need to see a clear leadership and decision-making structure in place when they consider an investment. And if this setup has had some time to bed in and prove that it works (the founders having worked together before), so much the better.

Only once you have these two in place, comes the merit of the idea, its disruptive potential, market size or opportunity for exit. No matter how ambitious a project is, if it does not have the right team, both in terms of commitment and team dynamics, it is not likely to get funded.

6

Coherent Labs

Company Profile

Product description: middleware technology and tools for game developers

Website: www.coherent-labs.com

Founded: 2012

Company size (# FTEs): 0–10 | 10–20 | **20–50** | 50–100 | 100+

Funds raised (in Euros): n.a. | 0–100,000 | 100,000–500,000 | **500,000–1,000,000** | 1,000,000+

Investors: LAUNCHub Ventures, Angel Investors

Annual revenue (in Euros): n.a. | 0–100,000 | 100,000–1,000,000 | **1,000,000+**

Interviewees: George Petrov, Founder and CEO, Coherent Labs; Todor Breshkov, Partner, LAUNCHub Ventures

V. Andonova et al., *Entrepreneurial Ecosystems in Unexpected Places*,
https://doi.org/10.1007/978-3-319-98219-9_6

Case Study

Coherent Labs was founded in 2012 by a team of four engineers. Prior to establishing and launching the company, the founding team was working for a single-game studio that managed to build and launch, but not successfully market and sell, its game.

Balanced teams, with one technical and one entrepreneurial founder, raise more money and have significantly higher user growth than purely technical or entrepreneurial founding teams (see Startup Genome Report version 1.1), making Coherent Labs the exception to the rule. Despite being founded by a team of only tech specialists, the company managed to attract early funding and successfully market and sell its product, reaching over US$1 million in revenue in 2016. In this case we will review the factors that helped an all-technical team become successful.

We have chosen to focus on this case in particular, because after investing and following the development of over 100 startups in the region, we have identified a general tendency for startups to focus primarily on product development while relatively neglecting customer development, sales and marketing. Engineers in the region are recognized for their skill and talent, and in many startups, the leading founder or the core or entire founding team is composed of engineers or developers. In a number of projects that were funded, the founders have the capacity and capability to execute even very difficult technical tasks, but have little or no experience with sales, and they often default to the 'we will build it and they (customers or users) will come' attitude to sales. In other cases, startups choose to maintain a one-way communication with their clients, either relying on social media messages, mailing campaigns or SEO, but rarely going out of their comfort zone or doing face-to-face sales meetings or customer interviews.

We have discussed this many times with both founders and investors, and we have concluded that there are two general factors that produce this behaviour: first, a general fear of failure among startup founders, and second, ego and pride, that still play a significant role in the life of startups. So, if the team is composed of very talented engineers, it is very hard for them to face the client, when there is a good chance that they will hear a 'no'.

In our analysis, there are two key factors that have played a positive role in the case of Coherent Labs in this respect. First, the team has had a long history together before launching the company, and, as noted by Todor Breshkov in his interview, there has been an established decision-making process and distribution of roles and functions within the team. Whether based on their previous experience with Masthead Studios or for some other reason, one of the founders has chosen to step up from his technical role and head the sales function of the company. So, regardless of the fact that all founders in the company came from a technical background, the sales role was recognized and adopted early on by one of the members of the core team.

Second, because the team was working in, and very passionate about gaming, they had first-hand experience of the problem they were aiming to address with their product and were confident that this problem applied not only to them but many other game developers globally. As George Petrov mentions in his interview, their initial marketing efforts through content marketing proved to be very successful because they were speaking the language of their users, and this allowed for frictionless SEO of their articles. And, as he said, after they received the first payment from a client, validating that they were actually creating value, it became easier to get to the second, third and subsequent deals.

Lessons Learned

Regardless of the background of the founders, there should be one key person taking charge of the sales and marketing function for a startup to be successful.

Founders' shared history, established roles and predictable team dynamics are instrumental for the development of a startup.

Founder Interview

Q. What were you doing before Coherent Labs? What made you choose to start your own startup? Tell us about the early days and how it all began.

A. Many people want to be entrepreneurs these days, but I have never been one of them—fixated on the idea that I should be an entrepreneur and work for myself. As a matter of fact, I have always felt very comfortable in every company I ever worked for before that, because I was lucky to be in a leadership or managerial position and I never missed what many entrepreneurs want—to have their own project and no one to tell them what to do and how to do it. So, from this perspective, I have always been an entrepreneur, although within other people's companies. What inspired me to start this company was a problem—the investor in the last company I worked for decided to stop funding it and it had to shut down. As my team and I really wanted to keep working in gaming, we had to establish Coherent Labs—out of necessity.

The company I worked for was called Masthead Studios. Before that I was working for Netage, which was a fantastic company, I was well established there, we had a great tech team and were working on exciting projects, but I have always wanted to be in gaming, being a gamer myself. Another thing I like about games is that they are in the technology—the pinnacle there—because you need virtually every other technology to build a good game. If you are dealing with financial software you just perfect some algorithms over and over; if you are in rendering you only focus on that; if you do AI, you do AI. Gaming combines all these things in a game if you want to make your NPCs (Non-Player Character) plausible and the gameplay intelligent. And on top of this, you need all these technologies to interact run-time. And finally, it must be fun. For these reasons, I have always been fascinated by gaming and wanted to work on something related.

Masthead Studios started with the extremely ambitious goal to create the first massively multi-player online game (MMO) in Bulgaria. At the time World of Warcraft was in its second year, was vastly popular, and everyone wanted to make MMOs. And when the opportunity presented itself, it was something almost unheard of—a funded gaming startup that lacked a key technology person. It couldn't get any better, so I took the chance and created the technology team of the company from scratch.

We chose to discarded the previous technology, and with very few exceptions to build all the technology for the game from scratch. It was very much my thing and we learned a great deal while working on it, but we also learned some hard lessons—games are hit driven, and no matter how good your technology team or your artists are, there is no way to tell if will be successful until you launch. Another thing we did wrong is a typical mistake for this region—we were building things, but not telling the world about it, so we failed at marketing the game and getting feedback. So, we launched the game as we wanted it, but the market did not accept some things and we could not fix them as we were out of runway. For example, we had the largest map in an MMO ever—256 square km—but it was a ghost world: you could walk for hours before meeting anyone else to play with.

Q. What made you choose this idea? What were your concerns at the time? What were the most difficult decisions and how did you approach taking them?

A. We burned through quite a lot of resources, the company grew to 115 people, but the game was not a success and the investor pulled out. One day the CEO of the company came to the office and gave us all a month's notice. It was just before GDC (Game Developers Conference), so together with three of my colleagues that later founded Coherent Labs we got on a plane and went to San Francisco for the conference. One thing led to another, and after a lot of elaboration on what we want to do, we figured we wanted to create a technology company in gaming—not a game, as that was too risky, but it had to be in gaming technology.

Then we started thinking about a couple of problems we had seen in the last five years that had been a pain for the team, and the user interface was definitely number one. There is a big gap between creative people and developers. Many systems in which there is a slow cycle between creation, implementation and feedback have an embedded conflict. And any such setup presents an opportunity for a solution. So we decided to make game user interface technology to solve exactly this. Then we tried to figure out which is the field with the most UI (User interface) and UX (User experience) people, and we realized those are the web technologies, and this is how we pinpointed the problem we wanted to solve, and that is how Coherent Labs was born.

As we came back from the United States we didn't have many resources, so we started some consulting projects and the four of us rented an office on the next floor, so we were spending eight hours on our day jobs, and after work we were working on Coherent Labs.

The next few months were hell. We were sleeping in the office and spending weekends there. In July, Eleven and LAUNCHub appeared on the scene and although we had defined our products, and we had secured almost a year of runway, we decided to go and speak with the two investment funds. It made some sense because we wanted to start travelling around and show our product to the world, and also, we needed to extend our team. Long story short, we pitched to both funds, got offers from both and decided to take an investment.

Q. What did it take to turn this idea into a business. What were the key challenges along the way? Did anyone help you with this?

A. What we wanted to do is focus on product development and coding, rather than getting involved in all the startup hype. In that we were lucky, and I don't think that if you were to build another startup the same way you would be successful. We were so focused on the product that we didn't think too deeply about the customers, but because we had been on the other side, and we were so passionate about it, when our product gets into the hands of other people like us, they become passionate as well.

And that is exactly how we created a business out of an idea. We have had this pain for a long time. We were building the product for ourselves.

Q. Tell us about your first hire and your first fire. How do you approach this, and how did it change from the time you started to today?

A. Corporate culture is one of the things that keeps me awake at night. It is not that we don't have it—the culture is as we want it, but it is not something that we have systemized appropriately. When the company was small, and I was doing all the interviews, everything worked fine. But we didn't have a structured company culture clearly described and defined, so at some point it became difficult to grow the team; there was a point when I fully delegated hiring and none of these hires is currently with the company.

All the big startup books say that the CEO's role is to build the team and set the corporate culture that is then channeled down through the company. However strange it is, companies, as an organism, copy their top-level team, and you can see this everywhere—in the culture, in the product and I can see that very clearly in Coherent Labs.

For example, in our company, all the developers are very passionate about the product, just like ourselves. And it is challenges that drive us. We love to work on projects of which other people say, 'that's impossible'. That is where all the fun starts, and how it all started for us—people said gaming is not compatible with the web—games need to be fast and the web browsers are slow and bloated. We still have that attitude, and when we see a non-web technology that is just as fast as or faster than ours, we will work hard to optimize and improve on our product. This is the culture that dissipates through the company, and when you are surrounded by like-minded people, passionate about games and technology, it works. Maybe this will not be applicable to just any business, as many of the people in our team are gamers themselves, and their work overlaps somewhat with their other interests, but we are lucky that this applies for our product.

Q. Tell us the story of your first customer. How did it happen?

A. The thing about our first client is, that I don't even know how they found us. It was a large mining company from Australia, and they were innovative in their approach to technology, as they were using game technology to make training simulations for their large excavation machines. So, one day, out of the blue, we received an email from the CTO of the training group of the company requesting a download of our tool, so that they could use web-based gaming simulations in their business.

Q. After that, how did you go from one user to ten? What comes next?

A. Once we got our first client, the first money landed in our bank account and we got that validation that we were doing something right and creating value for the clients, it became much easier to get to the second and third.

Our marketing strategy was set from day one, and as our founders are technical and not marketing, it was quite simply a very technical blog where we would write on how we solve particular problems—a topic interesting and understandable for our users. Because we were writing original content and named the things with their real names, we got good search results although we did not do much on SEO, we got quite a lot of direct hits from the right people.

Q. You've had several investments in your company. What do you think about investors? How do they help, and how do they hinder a startup? How about mentors—where did you find yours and how did they help?

A. We were again very lucky with investors. As we were not desperate to get investors and were not exactly certain we needed them, once we got an agreement with our first investor, we decided to push the limits and see what would happen—we asked for twice their maximum investment, as we figured we should raise enough for 18 months. Lyuben, the partner at LAUNCHub that we were talking with, liked our project a lot, and he committed to put his entire network to work on raising the additional capital we asked for, so we had our first round oversubscribed significantly and the luxury to choose which investors we wanted on board. This way, from day one, we had a good number of investors and mentors to work with.

My view on mentors has evolved significantly since the start and now I believe that a good mentor should be from the domain. The best mentors we have had are from gaming companies, and their advice is much more valuable than advice from investors. Investors are always ready to give you theoretical and rather general advice: you need to find the right people for the company. But that is not practical advice—that you can only get from people in the industry who have already done it themselves.

Investors bring two important things to the table—cash is one, the other is discipline. I believe it is very valuable when an investor disciplines the founders and challenges them. Another thing that worked well for us is that they made available a number of connections, who later helped us in various aspects.

I do not want to set unrealistic expectations here—to have a good mentor, he or she should come from your industry. For example, we needed help on structuring our pricing and got introduced to a lot of people to consult with, and the common advice was to move to SaaS. But SaaS is not something our industry likes, and if you follow blindly the advice of people that are not connected to your domain, you run the risk of taking a wrong decision.

Another thing I believe is really important is emotional support—to know that there is someone beside you, even if all he or she does is tell you that everything will be alright.

Q. What was your darkest hour? Have you had a 'near death' experience in your startup development? What was the reason? How did you overcome it?

A. At the end of our first year-and-a-bit we ran out of money, as many startups do. It was a bit different at the time, as we had more than US$70,000 in accounts receivable, but for various reasons I did not want to press our clients for payment. At the same time we had to pay for our booth at GDC for the next year, which was about US$50,000, so we didn't need all that much money, but our investors were not too supportive and told us we should go and collect on our dues instead of raising more money.

I had to borrow money personally, we stopped all payments to the founders for several months, sold half of our booth at GDC to another company and managed to keep the company afloat and make it to the conference. A few months later we landed several big deals, closed the accounts receivable and were back on track.

There was not a single moment when I considered shutting down the company, but there is always this feeling that our investors did not understand the situation fully and would not do what was to me the right thing. They may have been right, as we survived, and the company is stronger for this, but there was and still is this feeling of disappointment, that when you need money it is not available, but when you do not, everyone is willing to invest.

Q. What's next for Coherent Labs? What would you do if you sold your business? If you weren't doing Coherent Labs, what do you think would you be doing now?

A. Next for us is to change the way games are developed and change the way you see virtual reality on the web. Change the ways games are developed, because we are just taking off there, and we recently got in touch with some of the largest gaming companies in the world; they very much like what we are doing and there seems to be a good fit coming there.

On the VR—I myself am a big fan of VR. When Oculus Rift was launched on Kickstarter, I ordered one of the first devices, and it has been sitting around the office and we have been playing with it, running games on it and such. And when VR started picking up, companies and startups started contacting us, as it turned out we are uniquely positioned there, because everything on VR is made with game engines. But on the other hand, VR is a completely different concept of user interface, that is not available in games and the UI tools of games are very much unsuitable for 3D. We got really excited when several well-funded companies reached out and licensed our technology to apply in their VR developments. So, we gave it some thought and figured there is a very interesting intersect between VR and web—there is still very little content for VR as it is very expensive to build a big project on that technology, while there is huge amount of content on the web, and the web is getting more virtual and more 3D, and if the two could converge that would be amazing. This is where our concept emerged, and our idea on VR is to build a VR browser, allowing you to use all the functionality of 3D to present your web content. Our vision is that VR would be a game-changer for the web in the way mobile devices were back in the day.

However, this project requires a completely different type of browser technology in order to work, because standard web engines cannot render anything that is not in 2D. We started working on that, but soon decided to suspend the project, and the reason for that is that we are a startup. And if you are a startup, you cannot solve two separate problems at the same time. Focus is key, and there is a reason startups need to only focus on one project.

The really hard question here was, shall we focus only on the VR project, or do we go back to games middleware, and we had a hard time discussing this, but at the end of the day we are more or less established now with a team of forty people or so, so betting on a new project like that is quite risky, and that decision is not as easy as when you are just four founders, as we were at the start.

Q. What is it that makes you get out of bed every morning? How is your product making the world a better place?

A. I believe leadership means walking the walk, so just like anyone else in the company, I get up every day and come to do what needs to be done. But on a deeper level, I want to show the world that we are smart. You may call this ego. I am frustrated that in our part of the world when we start something we always do at a disadvantage I want to show the world that we are in no way worse than anybody else, if not better.

As for life after Coherent Labs, I have thought about it, and the feeling I have is, honestly, fear. I have this great thing going on right now, working with great people, and I fear that one day I will no longer have such meaning and purpose, after we exit the company.

Q. If there is one thing you know now, but wish you had known at the start, what would it be?

A. If I was to go back in time in the beginning, the one thing I would like to tell myself would be, create your own reality. Believe in the value of what you have created and demand a fair price, do not let the world shape what you are. I had shivers the first time I had to quote a price with four zeros after the digit. Now we are closing deals for half a million and feel comfortable about it.

Q. What was the role of the Bulgarian diaspora in Coherent Labs?

A. The diaspora was super helpful in what we did. We see the things here through a spyhole, while the people outside of this geography who had more experience than us could give us the bigger picture and are always available for advice. And being Bulgarian really helps, because the Bulgarian mentor is sincere about it—he is really interested in what we do. We have spoken with a lot of other people, but this level of relationship does not exist there—with the diaspora, these people really want to see you succeed, not just help you with one thing or another.

Q. What do you think would be different if you launched Coherent Labs out of Silicon Valley, for example, instead of Sofia?

A. It is quite a problem that we are doing this project out of Bulgaria—there is no one to discuss ideas with. We travelled to San Francisco at least three times last year—there is a weekly meetup on VR there, there are people who are excited about the idea. We made an incubator for VR startups and we had a hard time getting twenty people for it here, even though we were providing quite a good package for them.

And there is another thing about working here—Bulgaria is not yet a brand for entrepreneurial ecosystems, and people are naturally sceptical about teams or products coming from here. Simply, very few people know anything about Bulgaria.

As far as talent is concerned, there is again a big difference between the developers that we find here and the ones we meet in Silicon Valley. While we have very good, smart and talented engineers, their focus is on the code. If you speak with an engineer in a Silicon Valley company, he would be into the business of his company, the product and so on, while most people here are just producing pieces of code. Once again, we are very lucky to be in gaming and our engineers are passionate about games, so they would be seeing the big picture as well, but I am not sure this is true for most other companies.

Investor Interview

Q. What does 'investment readiness' mean for you as investors, and how did Coherent Labs fit into that?

A. It is very important to note that Coherent Labs was one of the first investments of our fund, and back then, both as a team and as a process we had developed the evaluation and selection process, but we had not tested it yet.

At that time, we were looking for several things in any project. Of course, a good team, which means a team that understands the product they are building and the market they are targeting, but also a team that possesses the skill set to handle most of the needs this project would have, at least in its early days.

In the case of Coherent Labs, we saw four founders that had a common history and background both with respect to their work together as a team, and with respect to the market and technology they were targeting. At the same time, the team had various skill sets that, as far as product development, product management and technology are concerned did not have any loopholes. Finally, the team had formed its internal hierarchy regarding leadership and decision making—there was one informal leader, which the rest of the team had recognized as such; for us it has always been important to understand how the team takes decisions and who is ultimately taking the decisions.

So, although there were deficiencies in the team in the fields of marketing and sales, for us Coherent Labs were investment ready. At this stage of our first fund we were looking for projects in acceleration and pre-seed phase. When we were considering them for investment, although they did not have a ready product, we did not think that was all that important. We thought they had enough experience and maturity to find and add the additional know-how to their team and project. In retrospect, this did not prove to be so easy, though.

Another thing that made us consider them investment ready was that they already had something they could demonstrate as technology and product at the time we were looking at the project. The team had spent quite some time previously working on this project full-time without any funding before they approached us. And I believe this is very important. We do not believe in people with good ideas who would go out and put together a team just because they have a chance to raise funding for it. But if this person or team has sacrificed and invested personal time in this project already, that demonstrates that they believe it has potential and they would do it either with us, or without us, one way or another.

Q. Is it important for the founders to have a history of working together?

A. Yes, it is. At this early stage, when you do not have a business yet, nor processes, and you are now going to hire new people, this shows that you have at least some basic level of team culture and understanding of the strengths and weaknesses of the people on the team, what motivates them and how you can work together. Those are elements that start to show in different forms in both difficult and happy situations.

A clear approach to leadership and decision making in a team is also important, because at the end of the day we are looking for teams, and there are situations where either an important, or a fast, or a difficult decision needs to be taken. Although there is no guarantee that this way you would always end up with the right decision, it is critical at this early stage not to get into a deadlock situation where nobody is willing or able to take a decision. We believe that this early in a startup, time is the most valuable resource, and it is important to be able to take decisions, and although some of these may not be the best decisions, the team would have the time to evaluate it, reconsider it, and take the next decision. So from this perspective it is very important to have within the team a developed understanding of how decisions are made, and who is ultimately making these decisions. We have had situations where we have declined investment in teams that claimed they always take decisions with consensus. In a team where every partner has equal weight in decision making, sometimes those partners have different backgrounds and agendas, and when they face difficult decisions sometimes they not only end up being incapable of taking any decision but become alienated from each other and leave the problem unsolved.

Q. What do you think is the founders' motivation to start this company?

A. I believe they have worked for several companies, one of which was funded in a similar manner, and at the same time having come across a certain problem many times, which they believe can be made into a product if addressed in the right way has motivated them to take this step and take the risk. It is a combination of a feeling you are an expert in the field and an ambition to do something more than a nine-to-five job.

People are driven and motivated by different things. People who start their own business are people who, consciously or not, want to solve some problem in the way they see it, and this drive is combined by some ambition, either fuelled by the desire to prove themselves, or to become rich, or to leave a trace behind, and to me none of these has any negative connotation. This motivation is even easier when you know that the labour market works in your favor—in particular developers and programmers, it surprised me that we do not see more of them trying to create their own businesses. At the end of the day they take almost no risk in today's labour market—if their project fails they can always get a job with some relevant experience under their belt.

I believe it is very positive for the environment in which we work to have a combination of branches of large technology companies, local companies that develop their own product successfully, as well as early-stage startups, although the latter would not always be successful. This sets an example of getting out there, and taking the risk, whether successfully or not.

At the start of our fund, after a few people from the large local tech companies joined our program, one of the founders of one such company was regarding us as a negative factor in the ecosystem, attracting the good employees and giving them a chance to develop a project on their own. Later, his opinion changed, once one of his employees quit, started a startup but did not succeed and joined the company back as a much more experienced and well-rounded professional than before. This happens because once you run your own business for a while, you get the full perspective on running a company and start appreciating things you took for granted before that.

And this recycling of startup entrepreneurs is certainly a substantial part of building a startup ecosystem. We see that the people that have made this sacrifice with their time, money, their family, and their opportunity cost, are much more full-bodied players in that ecosystem, but are also more demanding of the people with whom they work and communicate.

Q. How did Coherent Labs develop since their establishment in 2012?

A. One of the good things Coherent Labs did was that they launched a product on the market relatively quickly—not the perfect product, of course, but a product nonetheless. In the beginning they targeted smaller clients, but about a year down the road they realized that as long as they have a cheap product that targets the mass market, they cannot target enterprise clients. One of the difficult decisions they had to take early on was to discontinue the limited-functionality product. Although it was selling in large numbers, it sold at a very low price, and this hurt the positioning of the significantly more expensive enterprise product. So, at that point in time the team chose to focus exclusively on the enterprise version, at the expense of a portion of their revenue, without being certain they would be able to sell it just as well—after all, you sell to enterprises in a completely different way. In retrospect, it turned out to be a good decision, because even today, every year they go up the value chain of their clients—they sell to AAA titles, and the average license is about one hundred times higher than it was in the very beginning.

The other side of the coin is that this type of client requires a completely different sales organization, in which they partially succeeded, but partially are not there yet, because they still have a problem with the way they sell, with the sales cycle and with building the right processes and structures to sell in a stable, predictable and optimized way.

When it comes down to taking such hard decisions, it is ultimately the team that decides. As investors, our way of working is always to be close to the team in all such processes and to be a partner providing an additional perspective. It is not our role to take the decisions, but we are there to challenge them and encourage them to experiment more boldly, even at the cost of difficult decisions. The shift to enterprise sales is something we have discussed with the team a few times, but the final decision, its timing and implementation was something the team did. And certainly, we have not been the only ones with whom they have discussed and brainstormed this decision. We encouraged them, and all our other teams, early on to establish and maintain contacts with people who could be their advisors, mentors and partners. But always, and in the case of

Coherent Labs in particular, the ultimate decision is taken by the team, because they know their market, their clients, and the technology much better.

Q. What is one thing that Coherent Labs are better at than the other companies you have worked with?

A. It is hard to compare different companies, but within their market, one of the ways in which Coherent Labs surpass their competition is their technology and the way it works, and the value they create through their product for their clients. It is not by accident that they mostly work with large companies that could potentially become their clients. Whether they sell enough in terms of size and frequency, is a different matter.

The reason they are in a position to provide such a superior product and value is the excellent understanding of the technology and the development there. And this is one of the greatest advantages of Coherent Labs as a team. One of the reasons for that is that three out of the four founders are focused on the technology and the product. On the other hand, this is one of the setbacks, because as a team they still need to grow in other directions, with respect to sales.

Q. How do you see Coherent Labs as team?

A. They are not the type of team that is fascinated by the startup stories and legends, neither are they a team that loves going to startup events. They love to work on their product and develop the technology.

This puts them somewhat in a situation where they cannot fully take advantage of their environment and the potential amount of external help they could obtain. Just as an example, alongside our initial investment there were other investors close to similar products and markets as theirs, but they did not make the most of their access to these people, with respect to, for example, identifying and recruiting sales people, or how to certain processes in HR and marketing establish earlier or more efficiently. They had the opportunity to take advantage of this type of knowledge, but they did not fully take it.

Coherent Labs were one of the teams that showed us that it is normal to have different founders focus on different aspects of the startup. As there were four founders, it was very clearly delineated among them that some of them wanted to focus on the technical side, while others were willing to get out of their comfort zone and grow as management into different skill set, regardless if they liked it or not, but because someone had to do it. So, for us one of the learnings was that both are acceptable, but it is important for the founders to quickly figure out what they are willing to sacrifice or do, and not postpone such decisions for too long. If it is quickly clear who wants to do what, it is very easy for the team to figure out what additional resources and skill sets would be needed and what is already available.

Q. To what extent was Coherent Labs' progress affected by luck?

A. There is always luck involved in any successful startup. But it comes after a lot of hard work. In any company there is some degree of luck, and the big question is to what extent the team or person is capable of taking advantage of it, either by taking some risk or being more aggressive.

But work and luck go hand in hand—both are important. We have seen many teams that put a lot of effort in their work, but luck does not come, or it has come, but they have not spotted it. At the end of the day this is the combination that makes a successful entrepreneur—the awareness that there is some luck that you need to take advantage of.

Q. Looking back, is there one thing they could do better to be more successful?

A. One thing they could do is to be braver in taking advantage of their clients and more aggressive in establishing a physical presence in the American market. Maybe they were not as inclined to take risk in this direction. The big question most companies face when targeting the American market, especially when it comes to recruitment of people, is the difficult fact that you need to hire a person or a team which you are not certain you can manage and whose performance will be hard to measure, but at the same time you need to make a large initial investment in compensation. It is a chicken and egg problem—shall I first make a profit from sales, or fundraise and only then hire a team, or shall I risk six months of runway and get someone to establish me on this market first? And what if I am wrong?

And in most companies the cycle is pretty much the same: when they fundraise, instead of going into new markets aggressively (if this is where they are as a lifecycle), they postpone this step. This is a combination of, on the one hand, improper fundraising—if you had budgeted and communicated such a move in your fundraising process, then maybe it would have been logical to go after a larger round—and on the other, the difficulty of learning how to set your goals and what to look for in foreign-based sales people so you can feel comfortable you can properly motivate and manage them.

Q. What would have been different if they started in the Silicon Valley?

A. They would be much closer to their clients. Just like in any business—when you sell to large enterprise clients, it is much easier to sell if you are close to them, or other people they do business with, every day.

The other thing would have been the difference in the velocity and ambition of the environment. Things there operate at a different level of magnitude.

Q. How do you see the role of investors and advisors for the development of Coherent Labs?

A. I would like to say we did help. The first thing we do is to buy founders time, and try to help them to experiment more bravely, to learn from those experiments and apply this knowledge. Of course, as this is a mutual process, different teams can take advantage of that to a different level, and I do not mean that the ones that can better take advantage of this are more successful than the others. I even believe that Coherent Labs did not take the maximum resources neither from us nor from the other people around them, but this does not mean they are not successful. Each team has its own dynamics and perspective.

The positive with Coherent Labs is that we have always had a very easy-going relationship, as they are very open in communications and information.

The thing we could not help them with was to convince them that they should sit down more often to think and take a more ambitious approach to the projects. Our desire was to have achieved three times what they have now, but this is easier said than done.

7

Dronamics

Company Profile

Product description: unmanned cargo aircraft, capable of transporting up to 350 kg of payload at a range of 2500 kilometres

Website: www.dronamics.com

Founded: 2014

Company size (# FTEs): 0–10 | **10–20** | 20–50 | 50–100 | 100+

Funds raised (in Euros): n.a. | 0–100,000 | 100,000–500,000 | **500,000–1,000,000** | 1,000,000+

Investors: Eleven Ventures, Angel Investors

Annual revenue (in Euros): **n.a.** | 0–100,000 | 100,000–1,000,000 | 1,000,000+

Interviewees: Svilen Rangelov, Founder and CEO, Dronamics; Daniel Tomov, Partner, Eleven Ventures

V. Andonova et al., *Entrepreneurial Ecosystems in Unexpected Places*,
https://doi.org/10.1007/978-3-319-98219-9_7

Case Study

Dronamics was founded in 2014 by two brothers—Svilen and Konstantin Rangelov. Konstantin has the technical background, being a graduate of the Faculty of Aerospace Engineering in the Technical University of Delft, the Netherlands, while Svilen is the business co-founder with a background in business and finance. The purpose of the company is to build an unmanned, fuel-efficient cargo aeroplane (called Black Swan), capable of transporting 350 kg of goods over a distance of 2500 kilometres. With Black Swan the company is aiming to provide a valuable lifeline for getting goods in and out of small, remote, mountainous or island communities within hours at a cost below that of road transportation.

Dronamics have deliberately chosen to start their company in Bulgaria—a country with virtually no aerospace industry or traditions. In this case, we will present the pros and cons of starting a technological company in a market with little to no background in that domain.

Looking at the most critical factors for a startup to succeed, we can safely list the following:

(i) *Founders.* In early-stage startups, company culture is often equivalent to founder culture. To be able to motivate people, founders need to be driven, motivated and confident themselves.
(ii) *Team.* Founders are important, but they cannot accomplish much alone, so they need to find, hire, motivate and retain a competent team.
(iii) *Capital.* Runway is critical for innovative technological product-centric startups, where revenue comes only after a product is built and launched on the market.
(iv) *Advisors.* Whether you're building a complex product or not, advisors can help avoid a serious mistake or give you a solution right away that would otherwise take more time.
(v) *Access to partners and market.* When building a product that aims to overhaul established market practices, customer validation and early adjustments to the product are essential, and the way to do this is to talk to clients and partners early on.

In the case of Dronamics, only the first factor is obviously in place—Svilen and Konstantin have relevant backgrounds and are very passionate about the project. Access to capital, advisors and partners does not seem too challenging, as there is no daily need to interact with these, and it is possible to travel to meet the people involved.

So, what seems to be the most difficult task is finding a competent team, with a background in aerospace engineering, in a market with little to no aerospace industry. As it turns out, there are always some people with experience in any domain in any market, though not many, and they may be involved with something different at the time.

When launching in a saturated market, one must compete with established players and other early-stage startups for the same pool of talent, meaning it would be harder or more expensive to attract and retain the right people. But if there is only one player in a particular industry on a given market, people who are passionate about it would regard this as their number one choice, as they will have the chance to do something innovative and exciting.

For Dronamics, it took some time before the company had identified and hired the right team for their current stage of development. Initially they had several students from Konstantin's university come over to Sofia for a summer internship, and this has given a solid, but temporary, boost of progress. However, when the company became more popular in their circles, they started finding and attracting the right people.

Svilen identifies that, although at this early stage, as long as the team has up to 20 people, the lack of expertise is not so problematic, as the team needs to grow, finding people would become more difficult and the company would need to find and import talent from abroad or relocate to a more suitable market.

Finding clients and partners to validate and adjust the idea has been more challenging than expected, as there are few such companies in Bulgaria, and introductions are required abroad. However, the investors' network and the Bulgarian diaspora have been supportive with the necessary introductions.

Raising capital has proven to be the most challenging task for the project in its current market. The explanation for that, as given by Daniel Tomov, is the general attitude of European investors to disruptive innovation, where the investors take a more cautious approach and decisions are dependent on a plausible business plan. Furthermore, angel investors, and in many cases institutional investors, prefer to invest in their immediate market, both because this gives them easier oversight of the startups, and because they know the local legislation better and can work out the intricacies of the shareholding structure, corporate governance and control mechanisms. They also already have or develop specialization in a certain industry and tend to invest in projects in or adjacent to that industry. Accordingly, Dronamics' geographic location has posed challenges, primarily in the fundraising function.

Lessons Learned

There are both pros and cons for starting a technological company from a market with no traditions in the respective industry. In the very early stages, the advantages may outweigh the disadvantages.

Securing funding for such a startup would likely be the biggest challenge.

Scaling a company from such a market would be very difficult or impossible.

Founder Interview

Q. What were you doing before Dronamics? What made you choose to start your own startup? Tell us about the early days and how it all began.

A. For a long time, I have been interested in developing a business and applying things I have heard of or read about in a real-world environment. When I was in the tenth grade I enrolled in Junior Achievement and became the president of one of the student companies, where we were developing and selling calendars and other school materials. One time we even won JA's national business plan competition. A friend of mine had just returned from an exchange program abroad, and brought back a student planner/agenda, and we figured we could add advertising and distribute it to students for free—we were trying to make our classmates more organized, and we realized we could let advertisers pay for it. We later incorporated that into a business.

I have always wanted to be an entrepreneur, even when I was working for another company—especially right after 2008, when it was not the best time for entrepreneurship, so we were doing the agendas business after hours, while working from 9 to 5.

Looking at my career, I was a tax consultant at Ernst & Young for a year and a half, then another year and a half I was with a sports marketing agency, then I spent some time in a health food and urban logistics company: all in all, I worked to gain an understanding of all aspects of a business—financials, marketing, etc.

My drive to go into entrepreneurship was that my mother is a teacher and my father is a journalist—both professions rely on the government and its evaluation of their value. A state like Bulgaria does not value these professions very highly, and even today, both professions are not well paid, although they are key for society. So, I realized that you cannot always rely on society to put the proper value on the things that are important for it to function, and you would be better off looking after yourself, which you can do as an entrepreneur. Later on, when I was a consultant at EY, I got to know a bit about taxes and the social system, and it became obvious that the mathematics shows that if you rely purely on that, my generation would retire at 80, and with a miserable pension, so it is safer to be an entrepreneur, or to work for hi-tech companies where you can grow.

I analyzed the ways you can make money, and entrepreneurship is one, but it's not the only one—for example you can work for an investment bank and still walk away with a significant amount. For a while in college I actually wanted to be in investment banking, I was even heading the student investment club where we were trading US$200,000—more than some Bulgarian mutual funds at the time. However, by the time I graduated in 2007 what used to be one of the most desired professions was not so lucrative anymore, and all of a sudden it had all the risk of entrepreneurship but none of the upside. And you see, before everyone wanted to go to Wall Street, now everyone wants to go to San Francisco.

Q. What made you choose this idea? What were your concerns at the time? What were the most difficult decisions and how did you approach taking them?

A. The last thing I did before starting Dronamics was organizing large concerts. With a friend of mine we contacted the management of Jennifer Lopez and convinced them to add an extra date of her tour and start in Sofia, rather than in Belgrade. As we needed an investment for this, we landscaped the investment scene in Sofia and figured out what was available and what projects were fundable. We found an investor, the show was the most successful one in Bulgaria that year (2012) and overnight we became one of the largest agencies in Bulgaria. After a couple of years of successful shows I wanted to move on to something with more innovation and more impact.

Around that time, in 2013–14, my brother was studying aerospace engineering and working for an aerospace company in the Netherlands. Amazon drones had just been announced, and we were discussing them one time on Skype, working out that they would not work in our part of the world as many people here live in apartment buildings and a drone cannot deliver a parcel to the third floor and drop it on the balcony. Anyway, shortly after this conversation he asked if, when I was going to see him in the Netherlands, I could bring him some cheese from Bulgaria, and jokingly I suggested that he should just build a drone and have anything he wants delivered, and this is more or less how the idea of Dronamics was born.

We were circling around that idea, poking it from various angles, and realized that it made no sense to try to build a large plane, nor would it work if we aimed at small drones, as the former is very capital intensive, while the latter is very competitive. On the other hand, as the large markets are away from Bulgaria, it always reminded us that transportation for such peripheral geographies is very costly—you have at least two connecting flights to travel to the United States, or if you want to ship something to any large market you have to bear the extra shipping cost to stay competitive on price, so there is virtually a tax on being in the periphery as we were (and let's be honest, Bulgaria is not that peripheral, so imagine what it's like for some other countries). And although personally for me passenger transportation has been the bigger issue, we figured it will be much more difficult regulation-wise to build an unmanned passenger aircraft as compared to cargo. So, putting all these things together, we defined our product, which is an unmanned cargo aircraft that can transport 350 kilograms at up to 2500 kilometers.

One thing follows another; there was funding available for innovative startups, we had an idea that held water, so we decided to go for it, applied for financing and have been working on Dronamics ever since.

Q. Tell us about your first hire and your first fire. How do you approach this, and how did it change from the time you started to today?

A. When we started Dronamics, an accelerator was launched in my brother's university with similar terms to the ones we got in Bulgaria. We ended up choosing to start in Sofia, as the cost of living in the Netherlands would have given us a much shorter runway than we had here.

This choice limited us to relying mostly on Bulgarian engineers, and the good news here is that we still didn't need hundreds of people, so it is not all that difficult to find them, but eventually we expect we will hit that ceiling, which will be limiting. Bulgaria is 1/1000 of the world population, and the proportion of avionics or aerospace engineers is even smaller, which might be limiting our growth in the future. And the way to overcome that would be to either relocate or commit to bringing in talent from abroad.

We run into a problem here, because however much I would like it to be otherwise, the moment when someone would come home and tell his or her spouse 'Hey, honey, guess what?—we are moving to Bulgaria', and that would be an exciting thing, still hasn't come. The good news is that there is a growing expatriate community in Bulgaria, and once people move in, Sofia becomes one of the top three choices for these people, but we are still an undiscovered destination and there is more work to be done here.

When we started, in the summer of 2015, we had a number of people, mostly Bulgarians studying abroad, who came in for the summer break and helped us out with various tasks. Now we have a full-time team of twelve people, but we still have others with whom we work ad-hoc, as they are either not from Bulgaria, or are needed for a very specific part of the project.

For our company, being in Bulgaria cuts both ways—on the one hand, for people that are here and are interested in avionics we are one of the few, if not the only, option for them to work on something that is within their domain and is innovative and challenging. On the other hand, there are not too many people here, and in the long term we cannot rely only on bringing back people from abroad. In fact, half of our team are people that have studied abroad and have come back, but our funnel would be very small if we only stick with one nationality.

We often joke that if you draw a Venn diagram of investors, suppliers, talent, and customers for our project, its circles would barely touch Bulgaria. So, from this perspective we realize we cannot stay only in Bulgaria, but we plan to be abroad, on-site once we build the aircraft, so we can operate it within a larger airline in that geography. The target are mostly emerging markets, where e-commerce is outpacing the development of the transportation infrastructure, and people will need quick and cheap direct transport, as opposed to going through a central hub.

Q. Tell us the story of your first customer. How did it happen?

A. Our project is such that we cannot have a client before we are fully ready, and it is tough. As with every other company, when the end of the month comes you must make payroll. But with clients you also have a set of responsibilities, such as to maintain your services and continue providing a superior product, so even with client money, it is not always the case that revenue solves all problems. In our case, we just need to be more careful on how we use our funding money, time our funding rounds well and make sure we keep hitting our milestones.

Even though our product is not ready yet, we have interest from companies in Asia, Africa and the Middle East, and the more we validate the product, the more can we get this interest confirmed. At the end of the day, these companies believe in our capability to finish what we have started, and the prevailing concern is not so much about execution, as it is for regulation—again, something unique to this industry (software startups are hardly ever regulated).

In regulations we have a leg up, because we are very active in the emerging space of unmanned cargo transportation. Since this January we are the first strategic partner of the International Air Transport Association (IATA) for drones. We are often invited and awarded at conferences in our industry, and we have created a lot of relationships and recognition with customers and regulators. What we need to do now is to connect the dots: secure financing to enable execution and then get clients.

Q. You've had (a few) investments in your company. What do you think about investors? How do they help, and how do they hinder a startup? How about mentors—where did you find yours and how did they help?

A. My brother and I are the kind of people who are always keen to learn something new. With that mindset, financial investors can be very helpful beyond the money they provide, even if they do not intend to be. But being in our domain, it is hard to believe that anyone will invest in this project as yet another project in his or her portfolio, without getting involved in some additional way—providing contacts, bringing an outsider's point of view in a discussion, helping us with building the company itself, etc. And this sort of relationship with our investors enabled us to uncover additional resources that were very useful.

Having investors on board, we learned quite a bit about fundraising, and it also gave us validation to the outside world—people saw that someone was willing to bet his money on our project, which gave it extra credibility. We got access to a network of suppliers and advisors that later became investors.

When it comes to our mentors and advisors, they are mostly from the industry—scientists and engineers. They recognize that we are doing something quite difficult, but with a significant social impact, and are happy to be a part of it.

Q. What was your darkest hour? Have you had a 'near death' experience in your startup development? What was the reason? How did you overcome it?

A. The most difficult moment was probably about two years ago, when we ran out of funding—our fundraising process was delayed, as our milestone was delayed. In our case it is not a metric that you can partly achieve—as in, you need to hit 10,000 users, but you only managed 8000. In our project we either achieve it, or we do not. Eventually we managed, but we ran out of money before we met this milestone.

We had communicated very clearly to our team what was our runway, and everyone was aware that we only had funding till a certain date, and then we will have money again when we achieve this goal. I still remember the meeting with the team, when we announced we were at the end of our runway—we told each of them that we're out of money, but my brother and I will keep working to find investment. We told everyone that we can't pay them so we have to let them go, but as soon as we raise the funding, we would rehire them again, since we've been very happy to work with each of them. The next day, everyone showed up for work at 9 am, as if nothing had happened. And the next day, and the day after, and people stayed and worked without pay each day for the next three and a half months until we got the new funding. We were absolutely humbled, and every now and then someone would come up and ask 'Hey, can I leave earlier next Friday, I have to step out for an errand', and my brother and I would look at them and tell them we're not their bosses anymore, so of course they could. It is this team spirit that inspires me when it gets tough.

This was the most difficult moment, and this is still the most difficult subject, as we are dependent on funding. When it was just the two of us with my brother it was much easier, but now we are responsible for other people and whole families, so it is a thing we are constantly mindful of.

Q. What comes next for Dronamics? What would you do if/when you sell your business? If you weren't doing Dronamics, what do you think you would be doing now?

A. We are preparing for our next round of financing, as we just successfully tested a 1:4 prototype, including satellite control, meaning that the aircraft can fly and be controlled over a 2500-kilometer flight. With this funding we will build several full-size prototypes, so we can do customer trials as well. Next up would be building a facility where we can start manufacturing the aircraft.

Dronamics is not the end of the road for us—after cargo there would be great many things we can do in this space—it may be passengers, it may be another type of transportation. The world will be a completely different place in 10–15 years. There will be people on Mars, there will be a big heap of new problems to solve. So I don't think we will take a break after Dronamics, or if we do, it will be a short one.

Q. What is it that makes you get out of bed every morning? How is your product making the world a better place?

A. When I was studying economics, and even later, during my work, a topic that has always interested me was what people do with the money they earn. In a very oversimplified way, each of us gets an amount of money each month, or each week, what do we spend it on? On a basic level, we spend it on either things we enjoy, or things we don't like, but are necessary. No one is happy to pay their landlord or their mortgage. But when we spend cash to go to a concert, or to treat ourselves to a nice dinner, these expenditures are much smaller, but they are much more memorable, they're the ones we look forward to. Happiness value is not the same as monetary value. And then, there's a large part of the population of the world that has very little income to spend on anything, even necessities. So if there's a way to lift people out of poverty, you obviously increase their chance of increasing their happiness. Think of the so-called Taobao villages in China—entire villages and small towns in China that were able to lift themselves out of poverty just by learning how to sell their goods to consumers all over China by using Alibaba's marketplace Taobao. Those small artisans, those little mom-and-pop businesses in the countryside found a way to generate more revenue, leveraging existing technology. We asked ourselves, is there a way to lower costs and what technology would we need to create? And because the economy is so diverse, the one common problem all businesses with physical products have, is transportation. If the transportation infrastructure was better, shipping costs would decrease, and therefore the disposable income of those small entrepreneurs would also increase. Our drive is to solve this problem with Dronamics.

Another thing that helped us choose to do this project is a study that shows that, even in traditional businesses, 95% of startups fail during the first year, and by the fifth year this ratio is 99%. So, whether we are, for example, opening a restaurant, or whether we are aiming for as big a goal as ours, the difference in the risk is just a few percentage points, but the outcome would be much better in our case.

Q. If there is one thing you know now, but wish you had known at the start, what would it be?

A. I am getting more convinced, that we can do it in Bulgaria, and although it was difficult at the start, it will be getting easier going forward. We were less certain about that when we initially took the decision to start here.

Q. If you were to choose a different place to start Dronamics, where would that be and why?

A. The USA, Germany, China and maybe France. These are large countries with a well-developed aerospace sector, there is a lot of talent available and there is a history of financing such projects.

I think Dronamics would have looked rather different, if we had started in one of those countries with respect to, for example, the way we evaluate different risks. In any of these countries, with our current amount of funding we wouldn't have achieved even 20% of what we have achieved now, so probably our focus would have been primarily on fundraising. Another thing is, that in any of those markets there would be a better opportunity for state financing as well.

One of the differences between working on a small aircraft or working for one of the giants in the field is that you can spend years and years in, let's say, Boeing, working on a particular mechanism or, for example, the door of the aircraft. When working on a small plane you need to understand it in its entirety. This is where we need a lot of talent from abroad, as there is no specialization in Bulgaria for small plane design.

On the other hand, when talking about things like electronics, Bulgaria has abundant talent that is at par or better than in the rest of the world. As a matter of fact, there are some people we have met here that are very passionate about aeroplanes—for example we have met a guy that has built his own aeroplane from wood with a Trabant engine. And actually, Bulgaria has had some tradition in aerospace, and has been the first in a number of things in this domain: for example, King Ferdinand was the first head of state in the world to fly in an aeroplane in 1910, alongside a number of other engineering feats we have achieved, such as the world's first military airport (near Svilengrad, 1912), the world's first successful landing after engine failure (1912), and the world's first female military pilot (1912). And this curiosity and engineering ingenuity, combined with the scarceness of resources, is what makes a good fit for a product like ours, as for cargo it is not really important how the goods get from one point to another, but it's all about how cheap it is to do it, so we are thinking within this framework where we are always pushing down costs. Our goal is to build a cargo aeroplane that has a lower manufacturing cost and a lower operational cost than any aircraft in existence. Being in Bulgaria is definitely a large enabler for us to keep those costs down.

Q. Did the Bulgarian diaspora help and how?

A. I came back to Bulgaria in 2008, and there are many people abroad with whom I still keep in touch. And they have been very helpful. As the non-technical founder of Dronamics I was travelling a lot for conferences and meetings all around the world, and I have slept on many of their couches, but the thing they have been really useful with was an outsider's point of view, and contacts with potential investors and potential customers. Bulgaria is the country with the fastest declining population, but a lot of that is because alongside our population of 7 million, there are 2–3 million more Bulgarians who currently live outside of its borders. That is why the diaspora already does and will continue to play an ever-increasing role in Bulgaria's economy, and our company is an example where half of the team have already studied or worked abroad.

Investor Interview

Q. What does 'investment readiness' mean for you as investors, and how did Dronamics fit into that?

A. At the time of our investment in Dronamics, we defined investment readiness as having a good team, meaning on the one hand a good combination of people having some history together, and on the other hand having some business sense in the project you are working on. At this early stage in the development of a startup we considered Dronamics to have all the prerequisites for an investment from us.

There were several things. First, the team left a very good impression with us. The fact that the two founders are brothers, that carries certain dynamics, and they had very positive chemistry between them. One of them, Konstantin, carried the expertise with him, as he had worked on this in his university and the project was coming as a continuation of his study there, and he had the support of the university altogether. The other brother, Svilen, had a very interesting background in business and very good instincts about it. This appeared to be a very good combination of tech and business co-founders.

The other thing that was interesting about them is that their domain was in its infancy stage, and the project one of those wild bets that has a binary outcome—either something big happens or nothing at all. So, considering a domain that was just emerging, we believed that a few years later, as it was becoming more prevalent, they would be ahead of almost everyone else in the development of the aircraft and this will put them in a very good position.

From today's perspective, five or six years later, we want to see that people have invested more of themselves in what they are doing—either money, or time, or energy—but also to have something done, be it a prototype or something else, that is validated with the market. If we were considering an investment in Dronamics today, it would have been a tougher decision. And this is more a function of the growth of the entire ecosystem and the fact that there are more people that have invested themselves one way or another in what they are doing, giving you the opportunity to invest in more advanced teams.

Q. Is it important for the founders to have a history of working together?

A. In retrospect, it proves to be important. When you look at the situations where somewhere down the road contradictions arise within the team or some friction appears, and when you compare them with the teams that are steaming ahead, usually the basic difference is that the latter teams have had a history together before they started working on this project. They have been working together, have overcome various hardships together, they have adjusted their roles and personalities, and have learned how to unite and keep going in the face of these hardships rather than disintegrate.

It is not that this is always the case, but this is the basic thing that differentiates investment ready teams from all the rest.

Q. What is one thing that Dronamics are better at than the other companies you have worked with?

A. Dronamics stand out by being very dedicated—they had a number of challenges along the way, and although at each of those moments it seemed impossible to move on to the next stage, they managed to find a way. Whether it was to find the right people to work alongside with them—ranging from a person from the United States who used to work for Boeing to students from Delft, coming over for a summer internship, or whether it was to handle all the subcontractors here, because most of the parts were made to order and no one had ever made these before and they were looking through various alternatives to find the right ones. So, what I liked about them was that they were displaying character in all these situations, and not once did they come to us to say that something was not working, and they were considering dropping the project; instead, without ever complaining, they were constantly finding a way to overcome these challenges. And although many people told them how what they were doing was very tough and could not be done because they were in Bulgaria, with no traditions in aerospace engineering, within these four to five years, through conferences and meetings with a number of potential clients, they managed to position themselves as one of the leading ten teams worldwide in their space.

Q. What are the key qualities of founders that make them suitable for this task?

A. I think it is more of what I said earlier—being very dedicated and purposeful, not giving up in the face of a challenge, but entirely relying on yourself and being capable to build your company's story in order to attract other people to it. The latter is something that many founders could not achieve, as they did not manage to pull more people towards them, but kept to the original core team. Then when there was some hardship ahead they fought among each other and fell apart as a team.

With Dronamics, each challenge made them more united, and they keep pulling talented people around them, including some people who are much more experienced than they are.

Another thing is, that they are in an industry where there are virtually no institutional or other investors, and if there are, for example, four such investors around, they managed to convince all four that they can succeed in this project and despite the delays, because what they are doing is very hard and has not been done before, they hit milestone after milestone.

Q. What is the biggest challenge to the project?

A. There are a few things. The first is if they will manage to build this aircraft at all, although they made a prototype that can successfully take off and land and all systems that would operate the final version are tested and fully functional. But even if they build the full-size aircraft, the key challenge remains of how you would manage to integrate it into the existing business models in the industry, and deciding whether you would be a company that only sells intellectual property to larger companies, or try to establish a new business model on the market. The second option looks totally impossible from today's perspective, as it would require funding that no one has imagined they can raise. So, from today's perspective this is the biggest challenge: if a year from now they have a ready aircraft, in which direction would they be able to go from there?

Q. What was the most difficult moment for Dronamics?

A. They went through several tough moments. First, they changed the approach to constructing the aeroplane several times that set them back every time. The most difficult part of building an aeroplane from scratch, as opposed to adapting an existing one, is that you never know what your timeline would be. We spoke with their professor from Delft university, who is their key advisor, and who has, at some point, managed a team of over 500 engineers in Sikorski, and he explained that there is no blueprint for such a project—you cannot reasonably estimate timing and costs upfront. You may work on a specific part for some time and invest a lot of money in manufacturing it, but you will only know how it would behave once you have it, as computer simulations do not always predict precisely what is going to happen. So you may find yourself in a situation where you have invested a ton of money and time, and then you have to go back and start over again.

Second, several times they ran out of money—and this is always a tough moment, as you have to take a decision to go forward and maintain your team, while managing to convince investors to provide you additional runway, and explain to them that your original forecast understated your funding needs significantly.

Q. What would have been different if they started in a different market, and what would that market be?

A. If they started in the USA, maybe it would have been easier to raise more money to be able to get to where they are now faster, because the team made the whole project on budget. They tried to get on with some compromise on materials that actually delayed them and in the end it turned out they needed to use quality materials that require much more cash.

Companies in the same domain in the USA have started with initial rounds of US$10 million and above.

I do not think there are many other places in Europe that would have been much more beneficial. Maybe if they started in the Netherlands, somewhere near the university, that would have helped, but it wouldn't have solved the problem with financing. So, in summary, looking in retrospect the most promising places for this industry would be the USA and China.

In Europe, investors and partners assess companies like Dronamics on the basis of a very clearly defined business model and revenue generation strategy, and these should be very realistic and plausible. The same company in the USA will be funded because people understand that the world is changing, and although they don't know exactly how, they believe that somewhere along the way the founders will find the formula. In Europe, and especially in our region, conservatism and scepticism play a much bigger role. Looking beyond the current reality and believing in the creation of new business models is the American way, which doesn't work for Europe. And this is the reason the majority of ideas in Europe are either copycats, or smaller or larger optimizations of existing businesses or models, rather than an entirely new way of doing something.

Q. How do you see the role of investors and advisors in the development of Dronamics?

A. The investors' role in this project is that they have put their faith in the founders and their idea and have backed them up, and that to this day they trust them unconditionally and continue to support them. So, when the founders face some difficulties, you know that there is someone next to you, even if this someone has no specialist expertise in what you are doing, who gives you the confidence and reassurance that you can move forward.

I have seen this in other companies as well; even if you cannot help them with a particular solution, this gives them some confidence to keep going.

As far as mentors are concerned, in this particular project there were not too many mentors that could help them out. There were people that put them in touch with companies that could validate the need for their product, so the support here came more from the perspective of validation of the demand for this product or service.

If we are to compare this particular company, probably there are other markets where there would have been much more mentors available that could be of help to such a project. And actually the team found a number of mentors outside of Bulgaria, in the USA and in Europe.

Either way, founders that are too receptive to advice and always follow guidance are not always the most successful—founders need a certain level of stubbornness. There is a balance that can be hard to find between taking advice and acting on your own judgement.

Q. Did the Bulgarian diaspora play a role in the development of Dronamics and what makes Bulgarians abroad get more involved with Bulgarian startups?

A. Partially, yes. They had support from some Bulgarians working in American aerospace-related companies and there were some other people who are not in the industry, but are well connected and helped them with introductions and advice.

I haven't discussed the motivation of such people to get involved with local companies, but from what I see, I believe there is a level of nostalgia for Bulgaria, and some part of their lives which they cherish and they feel some sort of guilt not being there and not being able directly to influence and drive positive change, so when someone who is making something meaningful approaches them they try to help in any way they can.

8

Enhancv

Company Profile

Product description: automated builder for beautiful looking and organized resumes

Website: www.enhancv.com

Founded: 2014

Company size (# FTEs): 0–10 | **10–20** | 20–50 | 50–100 | 100+

Funds raised (in Euros): n.a. | 0–100,000 | 100,000–500,000 | **500,000–1,000,000** | 1,000,000+

Investors: Eleven Ventures, Angel Investors

Annual revenue (in Euros): n.a. | 0–100,000 | 100,000–1,000,000 | **1,000,000+**

Interviewees: Volen Vulkov, Founder and CEO, Enhancv; Svetlozar Georgiev, Angel Investor

V. Andonova et al., *Entrepreneurial Ecosystems in Unexpected Places*,
https://doi.org/10.1007/978-3-319-98219-9_8

Case Study

Enhancv was founded in 2014 by two friends, George and Volen, who were joined in the beginning by the third co-founder of the company, Dimitar. The company was building a product that helped job seekers create beautiful and well-organized CVs that would stand out from the crowd and get the candidate to an interview. The ultimate goal was to implement AI in the process that would give tips, hints and position-specific advice on how to edit the text of your CV so it will get through the computer filtering systems of the large HR agencies and companies and get you a slot on the HR manager's calendar.

There were two reasons that made us chose Enhancv as a case study for this book: first, that it is one of the few successful startups in the B2C space in the region and we wanted to analyse the factors that played a role in their explosive growth. Second, that at some point the company experienced a founder conflict, and we will take a look at their approach to team building and motivation and see how it worked and why in the end it did not work out.

The company was launched as a small-ticket investment by Eleven in 2014, and over the course of the next 18 months the fund invested up to its maximum permitted amount in the company over several tranches.

The team was quick to build and launch a product on the market, but for a while it was struggling with its conversion funnel, getting a good number of users, but not enough to break even given the low conversion ratios from free to paid.

By early 2016 the company was running out of cash and monthly revenues were in the range of couple of thousand euros, so not nearly enough to sustain the team. One of the founders, George, was fully engaged in fundraising and was trying to close a €300,000 round, but the process was very slow. And then something happened: in August 2016 the company booked US$35,000 of revenue, growing more than ten times over the previous month. In September it was close to US$50,000 and in October reached US$85,000.

At the time the company was experimenting with viral marketing—it was creating the CVs of prominent people or characters with their tool and posting them on social media. It turned out that their post of the resume of Marissa Mayer, then CEO of Yahoo!, got picked up by the media and social groups and went viral, getting thousands of shares and comments and a number of articles based on it. This brought huge traffic to the website of Enhancv and increased conversions, which coincided with the update of the pricing policy of the company, quickly taking it to over US$1 million of ARR.

The company has been trying ever since to replicate the success of the Marissa Mayer post, but with limited success, month-on-month growth slowing to single digits. The team is approaching the issue very systematically, analysing user behaviour and keeping track of metrics, but has come to realize that growth hacking is not something you can crack every time—sometimes you get lucky without being able to understand what the ingredients of this luck were.

Another area where the founding team of Enhancv is very focused and systematic is building a team spirit and culture—they were doing week-long brainstorming sprints away from the office, organizing retreats for the whole company, although some of the team members were working remotely, and even, at one point in time, attending Wing Chun training as a team. None of this, though, prevented the team from breaking apart and having one of the core founders leave the company three years after the start and a year after the big breakthrough with the Marissa Mayer post.

We have spoken with other founders about handling team dynamics and creating corporate culture, and none of them seemed to have a very committed and dedicated approach to that. In most cases it all came down to hiring the right people and leading by example, where the company culture was replicating the founders' and leaders' personal culture.

Lessons Learned

When running a startup, chance and luck will always play a role. What is important is to be prepared for a chance, to spot it, when it happens, and most of all, plan what you would do after that.

In a startup, company culture is not something you can create artificially: it comes from the founders and seeps down the organization. And while you are a small team, full honesty is key in building the right foundation for the company going forward.

Founder Interview

Q. What were you doing before Enhancv? What made you choose to start your own startup? Tell us about the early days and how it all began.

A. Before we started Enhancv I was a developer, but since we started I haven't done any actual development work. Coming from this background helped a great deal with Enhancv, and even with things like marketing, there turned out to be very interesting parallels between the two in the way you think about and approach problems and hack things. The need to stop and think, to break a problem into smaller parts and solve each of them is something that I learned from programming and later applied to things such as marketing. I believe that startups that have more developers on the team tend to be more successful exactly because of that—the way you approach and solve problems, rather than the ability to write code. Of the three founders at Enhancv, two have a background in product-centred thinking and that did help a lot.

Other than that, I may say that I took every opportunity that presented itself to try things out and play around. While at university, I went on an Erasmus programme. I was a part of a number of NGOs, I was also a part of a students' programme run by Microsoft, which gave me the opportunity to spend time in the United States. Overall, I took every opportunity I had to learn new things. The bad part was that after starting Enhancv, I had to leave all these activities behind, and it is always hard to stop doing something you love. So that was one of the hardest things when starting this startup—to learn to say 'no', to realize you don't have the resources to do many things simultaneously, and you need to focus on only one thing if you want to do it right.

When we started, it was me and my partner George, and a few weeks later we had Dimitar join the team. I always advise people with startup ideas to fill in an application to whichever accelerator or early-stage investor program they choose, to see their idea in perspective. At the time Eleven's application window was open, and that's exactly what we did. We didn't know much about entrepreneurship at the time (not that we know all that much now). We were part of the startup organization Start

It Smart (a pre-acceleration program), but this didn't bring much in terms of knowledge and expertise. And I am not saying that from the perspective of time—even in the early days we knew how very little we know. By the way, if you look at any Five Strengths, Gallup Tests and other such tools, everyone in our company is a learner. This is one of the main reasons people join and stay with the company—to learn. Sometimes achieving a result comes secondary to learning.

Looking back at the start, we were all learners, and again, one of the main reasons to start this company, other than believing in the idea, was that we were convinced we could learn a lot from it. When we started filling in the application form we understood how many things we didn't know, and I remember there were couple of months when we hardly slept, we spoke to many people, and we had this Google Doc with hundreds of versions and thousands of comments that we used to polish our thoughts. Little by little we began to understand how to put an application together and got a perspective on the idea. Again, if a startup approaches me today and tells me 'I have an idea', I'd advise them to give themselves a few weeks to fill in an application.

So, we had the idea, but very little more than that. We knew there was a problem, as we experienced it ourselves, and we had helped friends with their CVs, but we didn't understand the industry or the competition, or anything else for that matter. Our applying to the accelerator was the tool that helped us get a lot of clarity on the various aspects of the idea. And when I say application I mean not only the online questionnaire we had to fill in, but the process around getting the answers to those questions and probing them from different points of view, alongside the presentation we had to prepare later for the selection days that helped us get that much needed further clarity. We got accepted and that's how it all started for us.

Q. How did you and your core people meet? Is the whole team still together and what does it take to keep a team together for so long? If core team members moved on, what were the reasons and how did you handle it?

George and I were together at Start It Smart and he was one of my closest friends there. One day I went to him and said 'Listen, I have this idea, what do you think?' He liked it a lot, and the whole thing began. Start It Smart was at the office of Obecto Labs, a software development shop. We knew that the next thing we had to do was to recruit a tech person, so this was how we met Mitko, our third founder.

We didn't do much in terms of development and prototyping before getting into Eleven, but we had an online presence and some basic tools that helped us understand if someone would be willing to pay for such a service. We built a basic website with five resumes, and each of them had several buttons to look like it was automating the resume building process. There was a point in the journey where you were asked to input your credit card number, but after you did so it would show you a message: 'Sorry, but we are not processing payments yet.' We had maybe seven or eight people that got to this stage, and a few more people approached us directly and urged us to move this ahead, so we had at least some early validation of the idea.

To drive people to the site we wrote a small script that was parsing the job applications of one of the established job portals in Bulgaria with the emails of the job applicants, and then sent them an email urging them to use Enhancv to prepare their CVs. This helped us reach and drive a few thousand job seekers to the site, so we quickly established that these were our target customers, and in combination with the dummy payment service we figured that maybe people would be willing to pay for it.

Now that I think of it, building a CV and applying for a job is very similar to our process of preparing and applying to the accelerator. It helps you grow a lot, just like we did, and look at yourself through someone else's eyes.

When it comes to the team, we have parted with many people, and those have been predominantly our own mistakes. Our hiring today has little to do with the hiring back at the start. We learned a great deal about it, and I am not referring to just recruiting, but hiring as a whole. Nowadays it's quite difficult to get hired at Enhancv. Don't get me wrong, it's not easy to get quality applicants either, but we're much more careful in defining what exactly we're looking for, what this person is going to do, and once we've done that, the candidate needs to go through a number of culture fit tasks before we invite them to join the team. We would give them a number of tasks and see how they perform, then we would give them feedback and see how they responded to it. Overall, you need to give hiring time to make it actually work. But the most important thing we learned is not to compromise.

Someone leaving or firing someone is always hard, at least for me. Even people who have underperformed and who have clearly damaged the company, even with them it has been hard, and still is. It's like losing someone from your family. But as time passes it's getting a bit easier. We now see these partings as part of that person's journey—they spent some time with us, played their role, and maybe now is the time to move on. And if Enhancv gave that person the opportunity to understand what they want to do with their life—great! That means we've probably given them more than if they stayed with the company for another ten years. So, it gets a bit easier with time, but again, it's always quite hard. Just recently we had to part with one of the people that has been with us since almost the very beginning, but again, he has given the company what he could, the company has given him back what it could, and the relationship is rather exhausted. So it maybe it's better for both to part as friends instead of to stay and linger.

Q. What made you choose this idea? What were your concerns at the time? What were the most difficult decisions and how did you approach taking them?

A. Deciding on the idea behind Enhancv was a natural thing—we had this problem ourselves, and many of our friends had it too. We were at an age when everyone around us had a problem with building a recognizable CV.

And looking at the entire talent acquisition and hiring industry, the CV is the part of the process which most people get wrong. As an employer you cannot invite 100 out of 100 people for an interview, but there is a difference between inviting two and inviting fifteen. We figured that CVs are an integral part of the job search journey, and we wanted to make it easy to build an impressive CV, even if you don't have so much experience.

Maybe being naive and not knowing that much about this industry is what helped us not to despair and abandon the idea early on. In all honesty, we have stopped counting competition now, but we did not know about many of these companies at the start. There were many people that were telling us that there is no money in this business, and we couldn't monetize through subscription with a product like that, but we just kept going, as we saw we were making a difference, even if it was just for a few people at the time.

Q. What did it take to turn this idea into a business? What were the key challenges along the way? Did anyone help you with this?

A. We got an investment and started working hard on Enhancv in April 2014, but we didn't see significant revenue until August 2016, when a bunch of things fell into place, and we grew 100-fold over just a few weeks. But in the early days it was all about building the product, and when we got our first ever revenue we thought that there might be something in this idea.

But moving from idea to business took a while, and we got a lot of help along the way, even if some of the help came in the form of scepticism from our advisors, who were constantly telling us that there was a lot more ground to cover. That helped us a great deal, as was the case with one of our mentors, who kept pointing out things we didn't get right every time we met. He inspired us to focus on quality in every aspect of the product we were building and helped us realize the ways in which our product offers value.

Thinking back, we built a very niche product, which was solving just one problem—the problem with building a great CV. Compared to our competitors, however, our product was an order of magnitude better—not that we had some massive AI behind it, but it was just sleek, quick and easy to use. (Later on, we found out that our main competitor is actually Microsoft Word, but that is another story.) This was our first step towards making this a business—solving one problem very well.

The other thing we did was to focus heavily on marketing. The CV of Marissa Meyer we published was a cornerstone in the company's development, but there was a lot of preparation work done before that. We had decided that we will try to hack PR by publishing the CVs of famous people, and we had several such stories before that. But making the CV is not enough—you need someone to notice it and publish it. So, we started tracking the journalists that were covering our domain, we followed them on Twitter, studied their latest publications, we literally stalked them digitally before we decided how to approach them. There were days when I sent hundreds of very targeted personalized emails to these journalists and bloggers, trying to get through. It was a war—you try everything you can, and you aim very high all the time. There are many large companies that are not even trying to make high-level PR, while we were this tiny startup from Bulgaria that was trying to get the attention of top-notch US journalists. And eventually it worked.

Q. Tell us about your first hire and your first fire. How do you approach this, and how did it change from the time you started to today?

We spoke earlier about hiring and the way this process has evolved with us over time, but the way we do it now is very different from the way we did it in the early days. One thing we do now is publishing job offers. When we started, we never did that. We would meet someone, decide whether we liked them or not, and have them join the team if we did. But defining what this person would do and what is the best profile for this position came later. Now when we look for someone, we spend a good deal of time thinking about what factors would make this person successful at this job, and sometimes we go into minute details. For example, when we were looking for an operations person to handle administrative staff, we knew that we wanted someone who had graduated mathematics high school (a school type in Bulgaria), as this gives you a very thorough approach to your job; had lived or studied abroad, so they were open-minded; had worked for an NGO, as this is an indication you want to develop and grow. We had a list of seven or eight such traits that we believed would make this person fit for this particular position. Then we plugged all this into LinkedIn, it filtered down to three people, and we ended up hiring one of them.

The other thing we do now is based on the company values we defined and wrote down. It was another project we did, where we did a series of surveys and ended up with a list of values we believe in and we want to have in the company. Then we linked each value with certain behaviours, and from that we started thinking of interview questions that would help evaluate if this person shares our values and to what extent.

But even so, there is no guarantee that you will always end up with the right person on the first try. In the beginning we were trying to be a very fluid and a very consensual company—everyone had a veto on the hiring decisions. But this is no longer true. Sometimes there will be some people who don't agree with a particular hire. We will listen to everyone, have them express their opinion, but we won't be looking for consensus on every new team member. It is very hard to work this way, and as the company grows it gets harder to maintain this way of working.

Our first hire was also our first fire, but oddly enough, this person is still with the company, as we hired him back a while after we fired him. When we hired him, he was very young and this was his first job, so he didn't have much experience of working at all. Neither did we with working with people, so we allowed a number of exceptions to be made for him. This then led to people's motivation dropping, as they saw that and did not feel they were treated fairly. Things were clearly not working, and it was my responsibility to do the firing. This was probably the most difficult thing I have ever done, but I did it anyway. He was working remotely so I had to do it over the phone, which didn't make things any easier. Next thing you know, I got a call back from him and he said: 'Listen, you are right, I was wrong. I want to stay with the company, so tell me what to do and I am ready to do it. Even if I have to work for free, let us give it another try.' We agreed, and he is still part of the team. And we learned not to make exceptions—if there are rules, they apply to everyone in the company. Or course, there is some flexibility, but it must apply to all, equally.

But firing, just like hiring, is a process and you need to invest some time in it to do it right. You need to find the way to do it, to choose appropriate timing, to decide to what level you want to be transparent with the rest of the team, and at what point—not so much. Then you need to do an exit interview and off-board the person in the best possible way. There is no easy way to do it, and firing is always a hard thing to do, but it is an indispensable part of each company's life, and there are things you can do to make it easier and learn along the way.

Q. How do you plan to grow your business?

Growing ten-fold from where we stand today. This plan is comprised of marketing strategy and focus. I believed at some point that there was such a thing as growth hacking—you crack something and all of a sudden you grow tremendously. Then I realized there is no such thing. To have good marketing you need to know two things: why are people signing up to your service, and where are they coming from. I was just recently at an event where the speaker asked the audience how many of them know why people sign up for their products, and there were just a few hands raised. People really do not know why their users subscribe and what value they receive. We just recently did this research and realized that we also didn't know what our value proposition is. And when we found out and looked at our landing page we realized that we are not communicating this at all. So even if you hack some part of the journey, unless you know what your users are looking for, and what your value proposition is, you will fail them further down the road.

It's funny how many startups launch without giving marketing and sales its proper place. They say that marketing is something they will outsource and someone else will handle that, because it is something they don't understand. And they default to the trial and error method where they A/B test certain things but at the end of the day they are not quite certain why A worked better than B.

We were surprised to learn that to a large extent, our users sign up to us because they feel proud of the CV they produce. Sometimes they aren't even jobseekers at that point of time. We thought they signed up for the content, or for the tool, but it turned out they signed up because they can make a resume that makes them stand out, and once they have it they feel really proud of it and of themselves. They can submit it and say to themselves 'I did the best that I could. If I don't get this job, it's not because I didn't try hard enough.'

It's like in this example where when you buy a drill you are not actually buying the drill, but the holes you will make with it, and even further—the paintings you are going to hang on your wall once you drill the holes. So, buying a drill is actually buying the satisfaction of having these paintings hung on the wall—the drill is the tool for this end goal. Same with

Enhancv—it's not about building a resume, but being proud of your accomplishments and being able to present these in the best possible way.

Q. You've had (a few) investments in your company. What do you think about investors? How do they help, and how do they hinder a startup? How about mentors—where did you find yours and how did they help?

Fundraising is always tough. At one point of time in the life of the company we knew we needed to fundraise or die, so one of the founders focused entirely on fundraising and spent the better part of a year talking to investors. In the end, we managed to close a round for about a fifth of the money we were raising, but we started generating substantial revenue. And in all honesty, if this revenue had not happened, the round we raised would have been just enough to pay off the debts of the company. We literally had run out of our own and our parents' money, we had debts, and sometimes we couldn't even make it to payroll.

But it is not the investors that made this company. Don't get me wrong—all our investors are great and are very helpful, but they have not influenced the company in any of its major strategic decisions, or a strategic hire or a new investor lead.

For me, an investor should have a rather active role in the company. I am now looking at company decks for board meetings, and what Sequoya does is very impressive—they have a section in the deck on what the CEO did last quarter, but there is also a section on what the board has done for the company. It's not just a one-way street.

The most important thing an investor needs to bring to the company is an opinion. Yes, the company may take this opinion into consideration or not, but the investor needs to have an opinion. Whether this person has a broad knowledge of the company's business or a very narrow field of expertise is irrelevant—they need to bring their opinion to the table.

The other thing in regard to investors is their attitude. I once spoke to Connor Murphy of TechStars, and he told me that the relationship between investors and founders need not be a parent-child attitude. There is only one child there, and that's the company, and everyone needs to take care of the best interest of the company.

As far as mentors are concerned, we don't have dedicated mentors to work on every aspect of the business. When we face a challenge, we try to identify who would be the best person to speak to, and we try to find a way to get to them. And we are ready to listen and learn, we don't approach a meeting as if we already know the answers.

Q. What was your darkest hour? Have you had a 'near death' experience in your startup development? What was the reason? How did you overcome it?

The hardest moment came a few months ago, when we had to take a decision to part with one of the founders. It's not something that happened overnight but it was a process where we were drifting apart. We tried pretty much everything that we could to make this work and went as far as having sessions with psychologists to resolve this founder conflict, but nothing worked. And it is really very hard, having to part with one of the founding members—it's like breaking up with your first love. What is more, having a lot of people step in and take sides makes this even harder, and you start asking yourself if you are the good guy or the bad guy here, and it starts eating you up inside.

Q. What comes next for Enhancv? What would you do if/when you sell your business? If you weren't doing Enhancv, what do you think you would be doing now?

Right now, we have this founder issue and its repercussions to resolve before we can look far into the future, so that is our immediate task. The other, of course, is that now that we know this company can be successful and profitable, we want to focus on growth. But first things first.

While selling the company at some point in the future is on the agenda, it's not a very high priority now. And honestly, although I've given a lot of thought to what I would do if not Enhancv, or what I would do after it, I have no clear answers. I believe entrepreneurs are best recognized by their second project, not so much the first one. But whether I'll start a new project after Enhancv—I cannot tell yet.

Q. What is it that makes you get out of bed every morning? How is your product making the world a better place?

The ability to learn. That's the biggest part of it. To learn new things about the business, and new things about yourself, for that matter. To learn that sometimes it's acceptable to fail, and sometimes it's OK to lose some battles while other times you have to believe in yourself.

I would say that doing something, anything, is the easiest and best way to learn.

The other reason I get out of bed each morning are the people. When a new person joins the team, they receive something called 'The Five-Minute Journal'—a book where you log what you are grateful for each day. And many times, I would write that I am grateful for the fantastic team I work with.

Q. If there is one thing you know now, but wish you had known at the start, what would it be?

A. There are so many things we've learned that it is hard to pinpoint just one, but the one thing that really stands out is that speed and progress in a company are a matter of alignment. Meaning that everyone on the team knows the big picture, the final goal, and what is expected of them so that the company can get there. And it's very hard to align the team, if the company founders are not aligned. At Enhancv, many times it was the team that helped the three of us get that feeling of purpose and alignment.

But to get there, often you need to make some sacrifices. Everyone has their own opinion, and sometimes you need to follow a different path from the one you believe in, because you all set down together and decided to do so. But rules have to apply to everyone, and decisions, once taken, need to be followed by everyone. You have to swallow your ego and put your best foot forward, otherwise it won't work.

Q. What changed between the early days and the time after you crossed the 1 million euro revenue mark? How do you approach the project differently now, as opposed to before?

A. There are some things that I wish hadn't changed from the beginning, because I liked it better as it was before. For example, when we started, we lived and breathed Enhancv—we were working 24 hours a day. Now, if we do that, I know we will burn out, so if you look from outside it looks as if we are less committed, but actually we are much more productive this way. And the key difference is, that once you grow up, you start troubleshooting and problem solving, while at the beginning it is all about creating things and building stuff. And the latter is much more rewarding.

The other thing that came with the growth of the company is the responsibility. Not that we didn't feel responsible at the start, but now we have a larger team, more investors, and thousands of clients, whose lives you impact one way or another.

I remember one time we had a company retreat in Venice, and one of our teammates had a surprise for us. One of our clients, a gay girl, had an interesting story—she was coming from the banking industry and for a long time couldn't land a job, until she tried Enhancv and got a position at Spotify. So, before we started our working session we had her join over Skype, and tell us her story, and for me it was one of those 'Wow' moments—we are making a difference. I realized how very small things can have a huge impact on people's lives.

Q. What do you think would be different if you launched Enhancv out of Silicon Valley, for example, instead of Sofia?

I don't know where I would start and what would change if we were in the USA, but there is one key difference I notice between people here and people there, and it is their mentality and attitude—people here are very much effort driven and focused, while in the USA the results are what matters.

Being in such an environment would have probably changed our values, which may have helped us in being more disciplined, and probably being faster. But on the other hand, it feels like being in the army, which is not all that nice.

Investor Interview

Q. What does 'investment readiness' mean for you as investors, and how did Enhancv fit there?

A. The thing that made me invest in Enhancv was the vision and passion of George, one of the founders of the company, and the product from the perspective of creativity and utility. The utility and ease of use is what sells this product, and I saw the talent and skill set in Kaloyan, the company's designer, to execute these. But basically it was George, and his persistence—he was chasing me around at events and kept pitching and telling their story. It didn't happen overnight—it took a good three or four months, and it wasn't because I did not like the project, but I was not in active investment mode at the time.

As far as the product was concerned, at that stage I didn't evaluate its potential thoroughly—it was clear that it could be big, as it meets a fundamental need that pretty much every person in the world has on a recurring basis, because people change their jobs more often now. So, theoretically, there is a big market, but it was a startup project nonetheless—it could make it or fail.

Back then, and still today, I support and invest in people. Of course, if it's a stupid idea, no matter how good the people behind it are, I wouldn't back it up. So, it is people in the first place, their vision, and only then comes the lucrativeness of the project itself.

If I am to define and describe how I would rationally approach investment readiness, I would say it means to have a reasonable product-market fit and an understanding of how this will be exploited. On the other hand, it is about the team and their value system. But this is if I act rationally, and I don't necessarily always act based on an Excel spreadsheet, but rely to a large extent on gut feeling, so I am supporting people that I like, and who have a good vision and values. Sometimes the projects I invest in are more immature than they should be to be investment ready, but I give the teams the benefit of the doubt—they can learn along the way or they just need this impetus to try a second time instead of go to work for someone else.

Q. What is one thing that Enhancv are better at than the other companies you have worked with?

A. They have very good control and grasp of their metrics and are very diligent on measuring and tracking the behaviour of people.

Another thing they do well, is that they try to apply certain best practices, while other teams don't do that so much. For example, they hired an external company to help them analyze their pricing policy. And this is something, where you can be very wrong, and everybody thinks they understand it, but that is very much not the case. The fact that you're willing to spend tens of thousands of dollars on something so critical is a rather mature decision. In our company we have experimented with that, often with disastrous repercussions for our revenue.

They also worked with an adviser on their user experience analyses. So, overall, they aren't cutting corners on certain critical items, for which they don't have in-house expertise, and I regard this as very positive.

Q. How do you see the role of investors and advisors for the development of Enhancv?

A. I don't think that they will manage to take full advantage of the investors and advisers that are around them. Just as an example, the cost issues that they are currently trying to resolve have been there for more than a year, and none of their investors or advisers knew anything about it. They only came asking for help and advice once that thing was overboard, and the process was irreversible.

I think that, to a large extent, they fell victim to their success with revenue, which blinded them and they couldn't spot and address some fundamental processes in the company. Sometimes, by chance, you come across a product and the wave starts, but if a large part of the fundament is missing, you fail nevertheless.

Q. To what extent are startups affected by luck?

A. It does play a role. But chance only puts you in first gear, you have to then switch to second, third and so on, it cannot carry you all the way. So, sometimes luck comes, other times—not. Depending on timing, you may come across a very lucrative product and be very successful even with a less experienced team, or you may have the most stellar team ever, but still be run down by a tough market.

Q. How do you see Enhancv as team?

A. Enhancv have had a dedicated strategy for building and motivating the team, and I have often given them as an example in this aspect, but as it turns out, most of these things have been superficial. You cannot have team building days, pride yourself on your transparency and how everyone on the team is on the same page, to have one-to-one meetings all the time, and at the end it turns out that someone wasn't showing up for work, this person could not work with that person, and founders haven't spoken to each other for a year. So, fundamentally, they have tried to implement the right things, but evidently these didn't work.

It is critical in a team to maintain the relationships between the people. When there is a problem you sit down and talk. Because if there is a will, and if there is a team spirit, one can survive a crisis, while otherwise, when you start having gaps, even smaller things can irreparably damage a company. And it is about communication, but also about trust—no matter how much you communicate or how many one-to-ones you have, if you are hypocritical it is not going to work. It has to be honest. This takes a lot of work, honesty and compromise by each party, but when that is in place, things work.

9

FITE TV

Company Profile

Product description: a website, mobile application and streaming service for smart TVs, where users can discover and watch combat sports events on a pay-per-view basis

Website: www.fite.tv

Founded: 2012

Company size (# FTEs): 0–10 | **10–20** | 20–50 | 50–100 | 100+

Funds raised (in Euros): n.a. | 0–100,000 | 100,000–500,000 | 500,000–1,000,000 | **1,000,000+**

Investors: Earlybird Venture Capital, LAUNCHub Ventures, Angel Investors

Annual revenue (in Euros): n.a. | 0–100,000 | 100,000–1,000,000 | **1,000,000+**

Interviewees: Konstantin Yordanov, Founder and CEO, FITE TV; Dan Lupu, Partner, Earlybird Capital Ventures

V. Andonova et al., *Entrepreneurial Ecosystems in Unexpected Places*,
https://doi.org/10.1007/978-3-319-98219-9_9

Case Study

FITE was founded in 2011 as a spin-off of Bianor, one of the most successful Bulgarian service companies in the tech sector. FITE originally started as a product called iMediaShare, which was enabling casting of media content from a smartphone through the cloud to any connected TV.

We have picked FITE as a case study for this book, because they are a textbook example of a company that faces a roadblock in its development, but is smart enough to get back to the drawing board, scrutinize its assumptions and performance and take an informed and well-considered decision to pivot.

The company quickly gained significant traction and reached millions of downloads of its mobile app. Retention, however, was rather low, and the company took a look at the issue. As the founder, Konstantin, put it, they were trying to be more or less everything for everyone, and there were multiple possible use cases of the tool, but it was not perfect in any of them.

This made the team shift its concept under a new brand, Flipps, which not only allowed customers to cast their own content, but also provided a platform for video content. Soon after, the company figured that users were not satisfied with partial libraries; they expected to find just any movie they wanted to watch, and after a few unsuccessful searches, they gave up on the product. This led to low user retention and the team needed to revisit its strategy once again.

Finally, after some deliberation and careful analysis, the company did another pivot, narrowing down its focus from providing general media content to specialized media content in combat sports events under the brand FITE. The licencing model of combat sports events organizers quickly provided the company with access to a large pool of content, allowing their users to find if not all, then at least most of the shows they were looking for.

As the Startup Genome Report uncovered, startups that pivot once or twice raise 2.5 times more money, have 3.6 times better user growth and are 52% less likely to scale prematurely than startups that pivot more than twice or not at all. This is why the two pivots of FITE were what made us chose them for this case study, although it is a common trait of many startups to pivot in their attempts to find a good product–market fit. However, in other projects we have worked with, we have seen pivots, based on partial information or simply on the whim of the founder, so we want to make a difference between pivots that are well thought out and implemented and pivots for the pivot's sake. In the case of FITE, having investors on board made the company go the extra mile, as there was another person in the boardroom they needed to convince that they were turning in the right direction—as the saying goes, not many brilliant ideas were born in a boardroom, but many bad ones have died there.

Lessons Learned

Pivots have a positive role in the life of a startup, as long as they are based on data, carefully analysed (investors and advisors are a good sounding board here) and implemented.

As startups have limited resources, they can only do one thing extremely well. Trying to be a Swiss army knife is not something a startup can afford.

Founder Interview

Q. What were you doing before FITE TV/Bianor? What made you choose to start your own startup? Tell us about the early days and how it all began.

A. Bianor is the successor of the first company we started back in 1998. We did not have the idea to start a startup and build a specific project—it's as if it happened by chance. At the time my partner Nick and I were working at IBM, and the company decided to outsource the project we were working on. We liked this project, we spoke with the management and they agreed to outsource it to us if we started a company. So, our first startup was in a way the means to continue doing what we were doing before.

In 2000 we had our first exit, when we sold this business to Framfab, which was the largest internet consultancy in Europe. We started negotiating for the exit with another company that was looking to establish presence in Bulgaria, but it was acquired along the way by Framfab, and we ended up selling to them. This was the first and last time that we felt we made it big—there weren't many exits back then in Bulgaria, it was to a public company with a great culture, with a good deal of hype around it and great plans for the future.

Q. How did you and your core people meet? Is the whole team still together and what does it take to keep a team together for so long? If core team members moved on, what were the reasons and how did you handle it?

A. When in 2001 the Dot-com bubble burst, Framfab, being the largest consultancy in Europe, were most affected. We had the opportunity to buy back the company and as the plans were for the mother company to be taking care of the sales, we ended up with a 50-people engineering team with virtually no sales or sales channels. We changed the name of the company, the business model, and the domain, and with part of the team, because we could not keep everyone, we started Bianor.

While the first time around, in 1998, we had our own business and were by definition entrepreneurs, I believe that Bianor was actually our first startup. The mindset in the two businesses was completely different—with Bianor we had to start thinking about getting clients, struggled with making payroll every month and we actually faced all the challenges a startup faces.

Q. What made you choose this idea? What were your concerns at the time? What were the most difficult decisions and how did you approach taking them?

A. You rarely see startup spin-offs from established companies, and even less often you see ones that succeed. The reason we spun-off iMediaShare from Bianor was that we realized how different the startup culture is from what you see in an established company. Trying to do a startup-like project in a corporation usually fails rather quickly, and to a large extent the reason is exactly the difference in culture between the two. Large companies tend to work on minimizing risk, implementing processes and you have completely different goals overall. Startups on the other hand are quite different—it is all about getting things done, no one focuses on processes or risk management, especially in the early days. And if you start thinking corporate in a startup, you slow down, and with startups it is all about speed. The way I see startups is like this: you have some runway, and the key question you have is how much traction you can get within this runway to convince yourself, the market and the investors that it is worth more runway, which is a radically different mindset—it is about building quickly, making mistakes quickly and quickly learning from your mistakes. And the other thing you need in a startup is this heroic effort of working long hours and giving it your all, which is quite difficult to get in a corporate environment. I am not saying this is the standard mode of operation, even for a startup, but there are moments in the life of a startup when this is a must. It is a marathon, not a sprint, but there are times where you must sprint as fast as you can in the early days, as well as some moments later on, but you cannot sprint the whole way.

And of course, spinning-off helped get funding for that particular project, without having it mixed up with all the other corporate projects.

We figured all that out early on, when we were contemplating starting iMediaShare, and we got a team of dedicated people from Bianor that were fully committed only to this project.

Q. What did it take to turn this idea into a business? What were the key challenges along the way? Did anyone help you with this?

A. We started initially with a product called iMediaShare, then we moved to Flipps, which later evolved into FITE TV, and each of those steps has actually been moving to a more and more niche product. The first product, iMediaShare was streaming content from your phone to your smart TV—you could stream video, audio and photos. At some point in time we started adding our content to that—the product turned into a sort of Swiss army knife—it could do many things but was not perfect for any of its use cases. It was enabling streaming video, photo and audio content from your mobile phone, from the cloud or from social media, and each of these use cases demands a dedicated user experience. For example, if you are handling photos right you need well organized albums similar to how people organize their photos in the real world. Each of these cases requires a level of depth in the product to really add value to the users, and the bar is set very high in terms of user experience. You cannot compete with the level of user experience that the big players have if you are all over the place and you can only just about do the job. And if you are a startup you only have limited resources, with no means to cover a large number of use cases well, so you need to focus. So, it dawned on us that we needed to pick one segment and focus heavily on that.

We made many mistakes along the way, but we got one thing right from the start, and it was to integrate analytical tools in our apps, so that we could follow closely how our users behaved. Looking through this data we realized we were not happy with the level of user engagement, but also identified what our users were doing with our apps, and it was not always what we expected. Just as an example, some people were using the tool only to transfer their photos to their PlayStation.

It was frustrating, because we had so many great ideas, but not the resources to execute all of them. And while in the beginning we thought we were lacking in technical capacity, we quickly learned that the problem was more in the way we thought about the product. We were effectively trying to build seven different products with the resources of a startup. It was time to decide which of these seven use cases we wanted to focus on, and it was not an easy decision, because looking at the usage, it was not really clear which was the leading use case, and furthermore, the case we felt had the highest potential in the long run was not the one with the highest usage.

We ended up picking the case of streaming video content from the cloud to the television set, effectively building a virtual cable operator. Technically, it was the more challenging option as video is the most demanding media format of all, and securing quality of service is even more difficult, while at the same time you need to work with content providers, effectively becoming a marketplace, but this is where we saw the real potential to do something big and this was our first big step. We stripped a large part of the product out—we actually packaged the current product, and separated part of it into a new product, that was Flipps, and only focused on this use case.

The second step came a year and a half later, when we took the decision to niche even further. The reason this time was not so much product-related, as it was content-related. We realized once again that with the resources of a startup we couldn't aggregate any type of content well—we couldn't have music videos, movies, news, and everything else at a good enough level. Although we quickly signed about 250 content deals, we did not have the depth we needed for any individual type of content. We thought that we had nailed the user experience, as we had people referring friends and we were growing without focusing too much on user acquisition, but the long-term retention was low—there was not enough content depth. We quickly realized we couldn't compete with the companies with deep pockets for content acquisition. We were limited to revenue share transactions, and for example when we went to Disney, they loved the platform, and were ready to work with us, but would ask for a US$2 million minimum guarantee to make sure we meant business—at the end of the day they get approached by many startups and they can't work with just anyone.

We were again at a crossroads, and a conversation with one of our investors—Tim Draper, who is also an investor in Twitch—made us start thinking about where to go next. We already knew that we had to change something, and this was the conversation that triggered us to actually start doing it. Tim is a very experienced investor, and he knew better than to just go and tell us what to do. Instead, he told us the story of Justin TV, that later chose to focus on a niche, changed its brand to Twitch, became extremely successful in the e-sports niche, and were acquired by Amazon for close to a billion dollars.

So, we knew that to get higher user engagement, to build a loyal user base and start monetizing, we needed to pick a vertical and start going deeper into that vertical, instead of spreading too thin in many verticals. We looked at the key strengths of our technology—our ability to stream video directly from the cloud to TV screens. While this is not as important for news clips, in sports, for live streaming and three-hour shows, it is a must. We started looking at sports, and while sports are one of the most expensive content verticals as distribution rights are very expensive, and for sports like football deals are usually exclusive, we found that combat sports are not quite like that.

We did a very thorough analysis of three potential verticals in sports, ending up with a very serious document that we presented to our investors, as we felt that we needed to get everyone on board for this pivot. We were in fundraising mode, so not the best time to do a substantial pivot, but nevertheless, we had to do it. So, we picked three potential verticals and analyzed the positives and negatives of each one in depth, and the resources we needed to enter each of them. In parallel we started running experiments in our product to see which worked best, and combat sports performed extremely well both in user engagement and monetization.

We worked very hard to avoid personal biases while doing the analysis and experimenting, as often people know what they want in the very beginning and will simply look for evidence to substantiate their preference, but I must admit I am very happy that we chose combat sports as this is my chance to do two things I love—martial arts and technology—at the same time.

One of the advantages of combat sports is that the premium market there is all pay-per-view model based—you pay for each individual show and it has been like that for decades. The best quality content on the market is pay-per-view only, and even if you want to watch it on TV you have to pay your cable operator an additional fee. What is more, the established model is revenue share between the content providers and distributors and no distributor can get exclusive rights. For these reasons, it is a model that allows a startup to plug in—there are no minimum guarantees, the big guys cannot get exclusive rights, and you have access to premium content at par with everyone else.

Our pitch to content owners is very simple—we would tell them 'You already work with ten distributors, you need to start working with us as well. First, because we have a very targeted audience for exactly what you are providing, which is growing fast, and we have global reach, which none of your other distributors has; and second, because we are the only ones that would help you market your content through direct mail, push notifications, premium spot on the app, etc.'

Q. Tell us about your first hire and your first fire. How do you approach this, and how did it change from the time you started to today?

A. Unlike most startups, we do not like to focus too much on corporate culture, because when you try to create one, it becomes artificial. It is natural for a company of 100+ people to have a more structured and explicitly set corporate culture, but when you are a five to ten people company, the culture goes with the people, it is in the way we work, in everyone's personal involvement. This is where the personal example of the founders of the company plays a very big role. Everyone on the team witnesses each hire and fire, and this naturally shifts the corporate culture in one direction or another, setting an example of which traits we are looking for and which ones are not welcome.

We do not involve the entire team in new hires—it is usually three to four people that are involved with the process. We try to make sure we do not hire people that would totally not fit with the rest of the team, but there is no 'cultural committee' or another form of unanimous approval of each new hire. We have a very diverse team with some very senior industry people on our US team, so obviously they have a completely different culture from the product team here in Bulgaria, but we find a way to work with everyone. At the end of the day, for a startup it is the end goal that unites the team, and not a written down set of principles and values. And the best way to communicate that end goal and align all the people on the team with it is by working shoulder to shoulder—this is how you demonstrate the level of commitment, priorities, etc. While it is rather difficult to do this one-to-one with each person on the team, what we do is Happy Thursdays—this is an informal meeting of the team, where each Thursday we sit down and go over what we did last week, what we need to do next week and how this all fits in the big picture.

I recently spoke with our Google account managers, who told us about their corporate retreat that took place last year in Barcelona, and this year will happen in San Francisco or some other place like that. In contrast, our last team building event was a few months ago, when we covered one of the largest events in the world, and the entire team had to spend the whole night in the office, making sure everything went well. It was the second largest boxing match of 2017 and we were the exclusive digital distributor for the US, so stakes were very high. We had to take care of millions in revenues and tens of thousands of transactions happening within a few hours that required a huge level of customer support and quality of service monitoring, among others. All in all, it was very challenging, and the entire team was on the verge of exhaustion on the morning after, but at the same time it worked great for bonding the team, knowing that you can count on the person next to you in situations like that. Going through such a tough challenge and succeeding where sometimes even the largest industry players fail gives you a sense of accomplishment and breeds team spirit like nothing else.

Q. You've had (a few) investments in your company. What do you think about investors? How do they help, and how do they hinder a startup? How about mentors—where did you find yours and how did they help?

A. Our investors always do everything they can to help. This means contacts and introductions to other investors, and to corporations in our field. So, all along we have had external support, without having our investors get operationally involved or taking too much of our operational time.

As far as other entrepreneurs are concerned, we did not get as much direct support from them, and not because there is no desire to help, but there are only so many entrepreneurs in Bulgaria that have the experience that we needed. We usually need help on content or product-related issues, and there just are not that many companies that have gone along the whole road, so they can give you all the answers you need in an hour-long conversation. But everyone I have reached out to has responded, so the culture is there, but the depth of experience—not fully yet.

Q. What was your darkest hour? Have you had a 'near death' experience in your startup development? What was the reason? How did you overcome it?

A. The entire startup experience is a near death experience, or to be more accurate, a big rollercoaster. Literally, one day you organize and stream a multimillion-dollar event and you think how this is going to happen every week going forward, and the week after you lose a deal you worked on for nine months. So, there are huge ups and downs with each startup.

One of the things startup founders need to master is to be tougher in the face of these challenges and not give up on the downturns, but also not to lose their heads in the upswing. Another lesson we learned from our investors is that they do not focus all too much on such short-term peaks or valleys—they would keep their attention on the long term trends rather than on the immediate developments. Everything converges to an average line and the question is what the overall trajectory is.

Q. What is next for FITE TV? What would you do if/when you sell your business?

A. We have a strong feeling that we have a proper product-market fit, we have positive unit economics now, good user engagement and monetization, so things look bright on this side. The biggest risk we face is getting top content, as most of the revenues are concentrated in relatively few content providers, so we are to a large extent dependent on these relationships, so that is where we are focused now. We had our breakthroughs in this field, now we need to keep persisting. Our vision is to turn FITE TV into the destination place for combat sports fans and promoters—if you are interested in combat sports, this is the place you go to.

I have been an entrepreneur for so long, that I can't even think about going back to a corporate job. I do not see myself as fully dedicating myself to the community in Bulgaria once we exit FITE TV, but I would like to help with what I can, including sharing our knowledge and experience. As far as time allows, I am doing this even now—I am always open to meeting and exchanging experience with other startup founders. I believe that one of the strengths of the Silicon Valley is the openness of the players there and the belief that knowledge must flow back to the ecosystem.

Q. What is it that makes you get out of bed every morning? How is your product making the world a better place?

A. I would say curiosity and the desire to learn more. For me personally, this whole thing is a great learning experience. The amount of knowledge we have now of product development, customer acquisition, working with investors—all this is a very interesting personal experience for me. It's quite different from what I would have seen myself doing, as I am a technical person by background, so that makes it even more interesting.

And so is the entire startup experience as a whole—when I wake up in the morning I know that it will not be a boring day. It may be anything else, but not a boring day.

Q. If there is one thing you know now, but wish you had known at the start, what would it be?

A. There are many things, but if I have to pick one, it would be focus. On many levels. On the product level, on immediate goals and targets, but at any point in time, and on any level you need to be focused, as you have limited resources, and if you do not have focus you will not be able to make it through.

Q. What was the role of the Bulgarian diaspora in the life of FITE TV?

A. We have spent three and a half years in the San Francisco Bay Area before moving our headquarters to New York, as the majority of the combat sports activities happen there, so we have been exposed mostly to the diaspora in the Valley, and it has been great. There are regular meetups of the Bulgarian community in the Bay Area, and you get to share knowledge, get advice and network. You meet entrepreneurs, as well as people working at the large corporations there, like Google or Netflix, who are willing to share their expertise openly with you. Without having a specific milestone that we can say we hit specifically because of it, I can say that the diaspora was very helpful along the way for FITE TV.

And although I can say we are experienced and tough enough not to need to get comfort from feeling that we are not the only ones facing the challenges of running a startup, it is a nice feeling to meet and talk to other people doing the same thing and going through similar hardships.

Investor Interview

Q. What does 'investment readiness' mean for you as an investor, and how did FITE TV fit in there?

A. Just to put this into context, in terms of background I am mostly a B2B (Business-to-Business) investor. So for me, FITE was unusual in the beginning, and the reason I became interested is that I thought it is a B2C (Business-to-Consumer) play that will become successful based on the technology. They had unique technology that allowed users to cast video from the Cloud directly to connected TVs, and I thought at the time that was a compelling enough use case that would differentiate them, and help them get distribution deals which would deliver content under a different model from the prevailing high initial fees and minimum guarantees.

So, at the time this was my assumption, and it was supported by the high rate of growth in the number of users and I thought it was only a matter of time and a bit of focus to get additional content to grow the user base to a large enough number to become relevant for the big players.

This was the investment assumption, though what happened in practice is that things didn't work out that well. I think it was a combination of not pushing the use case enough, and the lack of specific content that would make it interesting for users, but growth in the user base plateaued because as the company was attracting large number of users, it was also churning a lot of users at the same time.

Kosta (one of the founders of the company) is a great guy, he is clearly very articulate, understands the space, is driven, has the vision for the company, was determined, and was in control of his team, and all these were very important to us. We were looking at a company coming out of Bulgaria and targeting a very large and very complicated market, which is media in the USA, but we thought Kosta has what it takes to drive the company and build a team around himself in the USA. It was also important that he was willing to personally relocate to the USA and create his team there.

He also had his partner in the USA and they had the team already, with which they had had a previous success building an outsourcing business.

Q. What is one thing that FITE does better than other companies you have invested in?

A. They have specific technology that is quite compelling and can drive a specific use case in a large market. I do not think they understood the market particularly well, nor do I think that we understood the market particularly well. We made the assumption that technology would drive a certain behaviour, which in the end did not pan out.

Our assumption was be that the casting technology would bring a lot of users, and then we would have enough users to be a big player, and if you are a content provider and what to access these users, this is how we want to work with you. Our goal was to get to five million users, but we didn't, because of low retention: people found the use case compelling initially, but you can get amazed at technology only for so long, it is not like it is going to drive you daily use.

Q. How did you feel about the company narrowing down its market twice—going from casting your own content, to general content provision to specialized content provision?

A. You need to be realistic and pragmatic about your investment function, and hopefully this is what we are doing. We thought that there were certain things they would be able to accomplish, and at some point it became obvious to us that they would not be able to accomplish them. And then the question was, do we still believe in the company, do we believe in the team, or do we throw in the towel. We had a case where we threw in the towel and stopped financing the company, but in this case we thought that they were driven and that they had a fighting chance.

In such a case the question is, if one thing doesn't work, where do we go from here? And they came up with this focus on combat sports which made sense to us.

Q. To what extent was the shift to combat sports based on analysis and data, and to what extent on gut feeling and chance?

A. In order to make an informed change in your business you have to be analytical and you have to be able to support your decision. At the end of the day it is not just your decision—you need to get buy-in from investors. Usually investors are rational people who need to hear a good argument in order to support decisions, especially when they involve additional financing, which was the case here. You need to be analytic, but again, you need to stumble upon an opportunity—it's not like opportunities are a dime a dozen, and you just have to pick from a checklist. You need to do your homework and you need to do identify and pursue certain opportunities, so you need to get a flair for getting past the initial hype and identifying something that could be an opportunity and start to investigate. So, you start with a hunch, but you have to develop a rational argument for it.

Q. How do you see the role of investors and advisors in the development of FITE?

A. The company, especially initially, needed resources to take this technology from a small company in Bulgaria and try to make it available in the USA. The resources they lacked the most were financial. And this is where we and LAUNCHub came in. Afterwards, I think the type of resources the company needed were a bit different, mostly in relationships, connections, and access to partners in the USA. Although we did provide introductions and were trying to help, obviously we were not the natural investors for FITE or for Flipps.

That is why we encouraged the company early on to try to fundraise from a media-focused venture capital fund in the USA. We tried this several times, but unfortunately it didn't work at reasonable terms, as the numbers and the evolution of the business did not support it. Our co-investors from the USA did help a little, but I don't think they were a major contributor either.

Q. How do you see FITE's team?

A. We were seeing the founders as being technically competent, but we had this thing where we were trying to push them to be quicker. We were not deeply involved with the wider team—I wouldn't be able to comment on the company culture per se.

Q. What is the thing that drives the team?

A. I think it is a personal thing in the end. Usually founders are driven, and I think Kosta is particularly driven—I don't know what specifically drives him, but I don't think it's money. I believe it's passion to build something the right way. And I think there is also a sense of responsibility towards his team and, very reassuring for us, his sense of responsibility to the investors that backed him.

Usually responsibility drives founders once their teams grow and they get invested, because what they are doing is not a one time thing, and your reputation follows you in your next thing.

Q. What would have been different if FITE had started in the Silicon Valley?

A. Most probably they would have raised from a more relevant fund. It's not only about access to investors, but about the visibility you can get early on, the connections you can form, the personal relationships that you develop and the cultural understanding.

There is, in the case of FITE, the naivety of an outsider who thinks they can become relevant in a very tough market. This is not inherently bad—it can either work or fail. We are still unclear how far this will work.

Q. Do you think the Bulgarian diaspora helped FITE?

A. I think there is always a noise around you, and you always have new contacts and people willing to help, but I don't think there was anything that made a difference. He had some contacts he only made because he was coming out of Bulgaria, but I don't think they had a major impact on the company.

10

Sensika Technologies

Company Profile

Product description: big data analytics and sentiment analysis service, allowing users to follow, analyse and manage web and social media content based on predefined search filters

Website: www.sensika.com

Founded: 2012

Company size (# FTEs): 0–10 | 10–20 | **20–50** | 50–100 | 100+

Funds raised (in Euros): n.a. | 0–100,000 | **100,000–500,000** | 500,000–1,000,000 | 1,000,000+

Investors: Eleven Ventures, Angel Investors

Annual revenue (in Euros): n.a. | 0–100,000 | 100,000–1,000,000 | **1,000,000+**

Interviewees: Konstantin Hristov, Founder and CEO, Sensika Technologies; Ivaylo Simov, Partner, Eleven Ventures

V. Andonova et al., *Entrepreneurial Ecosystems in Unexpected Places*,
https://doi.org/10.1007/978-3-319-98219-9_10

Case Study

Sensika was founded in 2011 by a team of senior IT professionals and soon after received an investment of €200,000 by Eleven Ventures. The company is building a big data analytics and sentiment analysis programme that crawls the web and social media; gives users a list of mentions of specific predefined keywords, analysing the tone of the posts; and allows for collaborative action on these.

As is the case with most companies we have worked with, Sensika launched first in Bulgaria as a test market, and after closing its first few accounts, targeted the Western European and US markets. After several years of struggle and very limited success in these markets, the company decided to shift its focus towards the Middle East and Asian markets, released a new version of its product capable of handling the relevant languages and shortly thereafter got its first clients in the UAE, steadily growing during the next year to over US$1 million of annual recurring revenue. This is the reason we have decided to include Sensika in this case study: they managed to successfully pivot their target geographical market after realizing they could not be successful in the obvious and seemingly natural local markets for their product.

Before becoming companies, startups are just a series of experiments, aiming to find product–market fit, or later on a repeatable growth pattern. We often advise our startups in the product development stage to write down a list of the three to five key assumptions that need to be true in order for their project to succeed. Oddly enough, many times we see startups invalidating each and every one of these assumptions through customer development or market research, but then still stick to their original concept and product. Even more surprisingly, sometimes startups would focus on building and optimizing their product without engaging in any customer development, investing a lot of time and resources in building the perfect product, just to realize later that this was not what their clients need, are willing to pay for or would buy from them.

Sensika did exactly that: when they started working on their product, the plan was to enter the European and US markets, and their product was initially supporting just the English language and was crawling this part of the web. This seems natural, when more than half of the web content is in English and when one of the two major groups of users uses English, these seem to be your natural markets.

The team spent almost a year building their product and optimizing it for speed and efficiency in order to be able to process large amounts of data in the fastest possible way. Only after they launched a full version, did they start going to conferences and talking to clients, initially in Western Europe, only to discover very quickly that their potential clients were already using one alternative or another of a product or service that solves the problem Sensika was solving. And although Sensika's solution was cheaper, faster and more up-to-date, for companies that have been working with a service provider for years on end, it was almost never an option to switch to some early-stage startup, especially one coming from Bulgaria, when they couldn't tell if they would be around six months later or would have failed and closed shop. As Konstantin explained in his interview, it was at a meeting at a conference when his potential client told him straight out that the chance for them to switch from their current provider to Sensika was slim to none.

It was a harsh realization and we have seen other startups give up and throw in the towel in a situation like that; after all, Sensika had just launched a technically superior product to what was available on the market, but were at the end of their cash with no deals on the horizon. And what is more, around that time, two of the core members of the team decided to quit Sensika to start their own startup. Most of the time this would certainly be the end of a startup, but not for Sensika.

Instead of giving up, or even worse persevering in trying to do what they just realized was very hard or impossible (it is surprising how many startups do exactly that), they took a sober look around and tried probing different markets. About a year earlier, the company had had an introduction to a potential investor in Dubai, and although the deal did not go through, they had spent a while entertaining the idea of entering the Middle East markets. So, when they were looking for new markets to try, the first choice was Turkey because of proximity and size and the second was the Middle East.

One of the original founders of Sensika who was not involved full-time with the project came back, fixing the team issue to a large extent and focusing on adapting the product to process the language, and Konstantin moved to Dubai to try to work the market from the inside. Shortly after that, the first deal was closed and more clients followed. Growth was slow, initially, but then it accelerated and brought the company to over US$1 million in revenue five years after the start.

When they started, Sensika thought as many startups do: there is this huge market (whatever the market is), and just capturing 1% of that market will make us wildly successful. And it almost never works. Simply the fact that there is a huge market does not mean there is a place for yet another startup there, primarily because of several factors: (i) it is a replacement game—you have to get your clients to replace an already existing and running solution with yours, and the cost of switching is usually prohibitive, and (ii) you are a startup—you may not be around a year later, so the risk your potential client is running in buying from you is larger than buying from an established company, regardless of the difference in price or performance. Sensika were smart enough to realize that, and instead of trying to enter a competitive market with established relationships, they entered a market that was too difficult for their competitors to entertain seriously—because of language or local customs and specifics. This is what made them successful.

Lessons Learned

Product or customer segment (e.g. enterprise vs. SME) pivots are natural and easy, but do not always work. A market pivot, particularly a huge shift such as Sensika made, is much harder, but going to a difficult market is sometimes the more feasible choice, because if it is difficult for you, it is probably difficult for everyone else.

Trying to capture a small fraction of a huge market almost never works for a startup. To replace an existing solution, you need to be an order of magnitude better than the present solutions. Instead, you should focus on opening a new market by either having an innovative product or going to a market that is yet unsaturated.

Founder Interview

Q. What were you doing before Sensika? What made you choose to start your own startup? Tell us about the early days and how it all began.

A. I have always had a very strong entrepreneurial spirit in my family. Even during socialist times, and despite having day jobs, my family has been involved with all sorts of business projects in order to make ends meet. In 1990 my father—an ordinary school teacher—managed in just five months to organize the manufacturing of canned fruit and vegetables and establish a partnership to export these to Japan along the Trans-Siberian Highway. The money he made he used to send me to study economics in a German university.

After graduation I worked for three years in an American company, where my job was to sell chemical technologies to various manufacturers, and this is where I learned how to sell.

After that I found myself at Citigroup and was assigned to a small IT project. The project was to build an encrypted mail system that was connecting with branches of other banks and was transferring information on payments made there for our clients. I had no clue about any of it, but I gave it a try, and it worked. This is where I realized the power of technology for the first time. This simple system turned our corporate bank with a staff of 40 into a powerful consumer bank with hundreds of street branches in a snap. And somewhere along the way I fell in love with technology and decided that this was the field I wanted to work in next.

My next job was at WizCom (now VMware), one of the largest outsourcing companies in Bulgaria, which was my real start in the tech field. I learned a great deal in this job, seeing how a couple of students from the American University in Bulgaria built a successful business from scratch that employed 200 developers at the time, and I learned what it means to be a global business, as the majority of our clients were in UK and USA. At WizCom I had the opportunity to travel and for the first time breathe the air of the Silicon Valley, and somewhere there, the idea of starting my own business was born. But I didn't consider starting a professional services company, instead I wanted to build products.

Q. How did you and your core people meet? Is the whole team still together and what does it take to keep a team together for so long? If core team members moved on, what were the reasons and how did you handle it?

A. After WizCom, and as I had no engineering background, I knew I have a great deal to learn before I get there, so next stop for me was a Bulgarian company that was developing out-of-the-box software and selling it out in Europe. This is where I met my first two partners, with one of whom I work at Sensika.

With them, we started working on our first startup, a product that is best described as Groupon over SMS: people would register their mobile numbers and shopping interests and receive discount offers via text message on items of interest. One of our partners produced a very efficient viral campaign that grew our database of registered users very quickly. At one point, for example, we had almost 10,000 people in Sofia interested in socks! We ended up with a large database, but when we started to sell, we realized that businesses in Bulgaria are not so keen on taking the risk of an innovative approach—they'd prefer to see that something is working before they adopt it themselves. We also made our first mistake. Instead of launching an anonymized search, giving access to the businesses to browse the contacts themselves and get a visual proof of the value we promised, I met with them and told them how many we have of this or that category, without actually handing them the tool. And they resisted. But what they told us was, that they do have their own databases, so if we could provide the tool to communicate with them they would buy it, so we quickly adapted our product into an Outlook for text messages. We quickly gained traction, sending hundreds of thousands of text messages a month. We got an investor, an office and our first employees. We were now technology entrepreneurs.

This business grew organically and with the cash it generated we started developing our next product—Favit—a centralized web experience, meaning that you had your place on the internet that aggregated the content streams of all your social accounts and RSS feeds, and tried to organize them by interest. We got an investment for this project, and so started yet another part of our journey as tech entrepreneurs and our learning curve, both for us, and for our investors as they were a startup themselves.

For example, after a successful launch and tens of thousands of users in the first month, we made our first big mistake—rather than focusing on upgrading and improving our current version, we started working on a new version of the product, and a few months later we launched version 2, which was only half ready, and extremely over-engineered.

At the same time, we had already given up 50% equity to our current investor, which left no room for other investors whatsoever.

So, long story short, a few years down the road, and after launching a few more half-finished versions of the tools, we started getting the signals from the market that our window of opportunity was closing—social media platforms started restricting the number of API calls per day, content curation platforms started to gain speed and traction. And although we knew what we need to do, we were all at the end of our emotional and creative charge, had depleted our financial resources, had exhausted the trust of our investors and were unable to attract any new investors because of the cap table.

Having run out of any options, we took a development project just to provide some more cash and runway to the company, and this project came from, as it turned out, the person who became our third partner at Sensika. He wanted us to develop a prototype of an external intelligence solution. This prototype work was the actual start of the Sensika project.

Q. What made you choose this idea? What were your concerns at the time? What were the most difficult decisions and how did you approach taking them?

A. As I said earlier, the idea chose us. We found ourselves with a ready prototype of a product that was supposed to be developed further by another group of people. But they bailed out for some reason even before the first line of code was written. Then, while we were writing the code for it, my partners and I got involved deeper and deeper, adding our point of view and ideas to it, so it gradually grew on us, and when we saw an opportunity to continue with it, we took it. It was a difficult decision, as it meant forfeiting our previous project, Favit, but we had run out of options there, and Sensika was a fresh start.

Q. What did it take to turn this idea into business? What were the key challenges along the way? Did anyone help you with this?

A. There are 4 jumps an entrepreneur has to make along the way to a success:

(i) take an idea and turn it into technology
(ii) the technology into a product
(iii) the product into a business
(iv) the business into a great company.

I co-created 4 products. The even ones got to the third stage, the odd ones to the second. I never had the chance to reach stage 4 yet.

Along the way great help, faith and support came from my family first, my partners and their families, our investors and clients.

Q. Tell us about your first hire and your first fire. How do you approach this, and how did it change from the time you started to today?

A. The first hire was not a hire per se, because it was the cousin of one of our partners in the first venture. We needed help and we looked within our families for it. This was back in 2006 and we are still working together. The first fire was dramatic: the guy sued us. Stressful times. I needed to hire an attorney, but eventually he won the case for us. It was such a relief. We became great friends with the lawyer.

Our approach to hiring is very straightforward: we hire for attitude. We are looking for people who are thirsty—the ones that want to work and get things done. And we are looking for people with integrity. And we do not have any sophisticated approach to identifying those qualities—over the years we have developed a very good sense about people, and when the three partners spend time at the interview, we quickly get a feeling if this person would fit in with us or not.

After graduating from Eleven (the accelerator venture fund), I was a mentor to a number of startups, and it is a great way to keep your finger on the pulse as far as people are concerned. Spending time with these young entrepreneurs gives you the opportunity to understand how they think and what drives them, and with their generation it is not money that works—it takes a completely different set of tools to motivate them and keep them happy. It also gives you access to great people, and as sometimes startups fail, or founders quit, you have the chance of having these people join your team, which is a very good shortcut to finding great people.

Q. Tell us the story of your first customer. How did it happen?

A. I recall the first client in Sensika—Vivacom, the largest telco in Bulgaria. We had just come out of Eleven. The accelerator staged regular demo days in their office in downtown Sofia. Interesting days. The startups pitched in front of potential investors and clients. It was a great school.

So, I get out, pitch, get down from the stage, mingle. At some point Ivo, one of the three managing partners at Eleven, grabs me by the hand and leads me to a very calm gentleman that was sitting discreetly at the back. This is Nasko Dobrev, he says. I get a business card, he gets a business card. Mine says—CEO of Sensika. His said –CEO of Vivacom. I thought, wow! Mr. Dobrev said: 'It was a nice presentation. I don't have any idea what your product does, but somehow, I think that we need it.' I didn't know how to feel after that punch line.

One week later we got a meeting with the Head of Communications at Vivacom—Mrs Vesselka Voutkova. I still remember that October in 2013 when I got a call from Vesselka and she said that they decided to go with Sensika. The whole company (like seven people) went out for lunch, we had wine. That memory is so sweet.

Q. After that, how did you go from 1 user to 10? And from 10 to 100? What comes next?

A. Once you have a product, the only thing you have to do is push sales. Back in the day I was the only one doing it. I had a target to cold call twenty potential customers per week. Also, agents worked very well for us, people who open doors to members of management boards.

We collected all the contacts that we could think of and tried to get them to open doors for us on a commission basis. This also got clients. This is how the second largest telco in Bulgaria became a client—with the help of a door opener.

At some point we started to get contacts by word of mouth. Currently we have 88 clients in Europe and the Middle East. Enterprises, agencies and governments. Insh'allah, we'll reach 100 clients soon. But we think of another figure now—thousands. And that has to do with our product expansion plans, but more on that later.

Q. You've had (a few) investments in your company. What do you think about investors? How do they help, and how do they hinder a startup? How about mentors—where did you find yours and how did they help?

A. When you start a technology startup, it is as if your entire team sinks down to the bottom of the ocean and starts writing code. And this is where investors come in—they provide the oxygen you need in the early days, before you break the surface, and start getting money from your clients rather than from investors.

But there is more to investors than just money. We have two institutional and four angel investors in Sensika. Going through Eleven's acceleration programme was a very steep learning curve for me—I probably learned more than I learned throughout all my time at university.

Our first investor, Eleven, introduced us to our first client, and later to our second institutional investor, BIG. Our angel investors from the Middle East helped tremendously with business development in the region, opening the right doors and pointing us to the right people. When investors groom a young company, they help. They create opportunities.

But sometimes investors can hurt the business as well. If the investor does not understand the details of a startup's business, they can be less instrumental or even harmful, despite their good intentions.

Q. What was your darkest hour? Have you had a 'near death' experience in your startup development? What was the reason? How did you overcome it?

A. The worst moment for Sensika came in 2014, when one of our partners left the company. It wasn't a civilized divorce, as on his last day this guy stood in front of the entire team and declared that our product was crap. And he was the guy in charge of product development, so although it did not make much sense, it brought doubt and fear to the entire team. Moreover, we had been with him for a long time and losing a person like this makes you start questioning your judgement, and your common sense altogether.

But I have this quality, where I am unable to spend more than five minutes in a state of despair. My mind starts looking for solutions and starts considering not only the negative aspects of this, but also the opportunities. So, when this happened our fourth partner, who was somewhat removed from Sensika, came back on board with full speed, he was literally living in the office and working 24 hours a day. His enthusiasm and energy quickly spread to the rest of the team and everyone was back on board in very short time. Two and a half months later we had a new version of the product and the company was in better shape than ever.

Other than this, we had two times when were completely out of cash. We missed payroll for a month. We delayed paying taxes. The tax authorities were all over us, blocking accounts and demanding payments. Those were tough times, but we persevered and survived.

Q. What comes next for Sensika? What would you do if/when you sell your business? If you weren't doing Sensika, what do you think you would be doing now?

A. Launching our next version with self-purchase, self-service. We will empower smaller businesses with actionable insights. This will be the new search.

Q. What is it that makes you get out of bed every morning? How is your product making the world a better place?

A. The alarm and the need for coffee.

But really, I remember the last episode of *Breaking Bad*, when Walter White faced his wife and rather than repeating his usual 'I do it for you and for the family', he said how he has always done it for himself—it is something he is good at that makes him feel alive, so all along he was doing it for himself. That is how I think of it—I am good at this and this makes me feel accomplished, because this is what I can do.

In the world where data is the new oil, we give businesses the energy, so they move forward, develop and succeed. It pretty much gets down to who utilizes information better and we turned Sensika into a major weapon in this.

Q. If there is one thing you know now, but wish you had known at the start, what would it be?

A. Entrepreneurship and family don't mix well. I didn't know how to balance the two and I was spending so much time, effort and emotion with the business, that I had none left for my family, so at a certain point in time I lost them. How can I have both a happy family and a successful business? I lost the first by trying to get the latter.

Q. How did you end up targeting the Middle East? What were your major concerns and what were the key driving forces that made you choose that market?

A. We had been running Sensika for some time, got our first clients in Bulgaria—Vivacom and a few more. From the very beginning we knew that Bulgaria is a very small market, and we felt ready to take it global and try to sell abroad.

I remember we were at this business conference in Berlin and we met and spoke with many people, but one conversation stood out. One of the people I met there was working for one of our target customers, and as I was pitching Sensika to her, she said very clearly, that they have been working with their service provider for more than ten years, and she although she liked the product, there was nothing in the world, not price, nor features nor product quality, that would make her company replace their existing service provider with a tiny startup from Bulgaria.

That did startle us, and we looked around, just to realize, that the companies that need the service our product provides already had their service providers and have been with them for a long time. B2B sales are not like B2C, where there is much less loyalty and switching costs. So we figured that the markets we were originally eyeing—Western Europe and the United States—would be very difficult to enter and would require huge resources.

So, we came back and started thinking of what we could do next. We looked for markets that are not as saturated, and Turkey was a natural choice, as it was just across the border, and there is a large Turkish-speaking population in Bulgaria, so it was easy to recruit smart people to test and improve the algorithms.

Additionally, my family has a history of working in the Middle East: my grandfather used to work on construction projects there, so he was telling me the stories how technologies were slow to enter this market.

Lucky chance, around that time Eleven introduced us to a business development person in Dubai, who was representing an investment fund that was looking for potential investment opportunities in CEE.

Next thing you know, I was in Dubai and one of my partners was in Turkey, landscaping the two markets to try to find the next place for Sensika. The Arab world was a greenfield—because of language issues the players in our markets had not ventured there yet, and at the same time we felt close to the way business is done there. We have been involved in helping Syrian refugees settle down in Bulgaria, so we had access to the right Arabic speakers, and just recently one of the low-cost airlines had launched a direct flight to Dubai. We spoke with our investors and they were sincerely intrigued by the opportunity, so the decision was made.

It was very important for us this time around to get clients before we started working on the version of the product, as it was a significant update and would take a lot of resources which we barely had. So, I was spending one week a month in Dubai, talking to clients and selling, while back in Sofia we were working on closing the next funding round for the company. We got two angel investors—Bulgarians who lived in Dubai, alongside some of our current investors, and this helped open many doors and eventually close our fist deal in Dubai, with the government of Dubai. Today, three years later, there are just a few days left before, for the third year in a row, Sensika will be the exclusive media analytics platform for one of the largest events held there—the World Government Summit.

Now, when we are looking at ways to grow Sensika and which new markets to enter, we would do a thorough analysis and research before we take a decision, but back then we had neither the resources nor the time to do that, so it was a leap of faith as much as it was a gut feeling and alignment of circumstances. But I believe we made the right choice.

Q. What was the role of the Bulgarian diaspora in the development of the company?

A. There are just around seven thousand Bulgarians registered in Dubai, so there is no huge diaspora there, but the several Bulgarians we met and have worked with did help a great deal—with tips, introductions and opening doors. This is how we got our first clients there, and some of these people are now our investors, so even in a place as unlikely as Dubai, the diaspora played a big role.

I use this opportunity to tell anyone making plans to work in the Middle East that they can reach out anytime, and I will be more than happy to give back and help as I can.

Investor Interview

Q. What does 'investment readiness' mean for you as an investor, and how did Sensika Technologies fit into that?

A. Sensika are a good example of factors that would be a compelling investment case even for our current fund, which is targeting more advanced projects. Back then they were a more senior team that had a lot of prior experience and domain expertise. They also managed to sell us a specific use case their technology could solve. So, they knew what they were doing, had a reasonable pitch and had the team to back it up.

There were certain things that emerged later that slowed them down: the initial team composition was a bit odd, in the beginning—one or two of the founders were not fully involved in the project, and there was a person on the team that was supposed to handle sales, but that didn't manage so well. There were some mistakes in the initial setup, but the essence of a team with knowledge and understanding in their domain was in place.

Their common history working as a team was important, as we have seen many times that teams put together just for the purpose of starting a project fall apart. However, sometimes having common history can cut both ways—when people have been working together for too long, that brings tension and is a prerequisite for tension and separation. In their case it was an advantage that they knew each other from before, but this did not stop two of the founding members of the team quitting a couple of years later.

And this is where there has to be a balance; if some people believe they are carrying more weight, but have a smaller share (and obviously in the case of Sensika the shares were distributed oddly as people on the team that were not fully involved had a disproportionately large number of shares) that was a recipe for friction and eventually a breakdown.

On the other hand, one thing that worked well in this case is that Vassil Velichkov, the founder that was detached from the operations of the company in the beginning, was able at that moment of crisis to come back full-time and this gave a boost to the project's development.

Q. Is there something Sensika is much better at than the other teams you have invested in?

A. One thing they did, which very few other teams manage to do, is that when two of their key people left the team they managed to recover. The ability to build a team is a make-or-break for the startup—there are some startups that can recruit and retain people and there are others that cannot. It is even more difficult, when two of the core people of a team quit, to replace them, and get out of such a situation stronger, especially given the limited resources a startup has at its disposal. So this is something where Sensika did succeed, when many other teams have failed.

Another thing they did, for which we criticized them more than praised them, was that they were focused on the performance of their software and its scalability—we always perceived that as putting the cart before the horse. It goes against the grain of startup thought, but the team was committed to applying it early on and maybe it was something that helped them down the road. After all, it is an advantage that they can demonstrate to their clients that they use such advanced technology and that it has such a quick response times.

Q. What do you think about their pivot to the Middle East and was it serendipitous, or a well thought-out and analyzed plan?

A. As far as I remember, there was an investor from the Middle East that approached us, and asked for an introduction to several of our teams, and Sensika were one of them. Maybe this contributed somewhat to their turning their attention to the Middle East as a potential market, but they were the ones that spotted this was an opportunity. When we made the introduction, we considered this as a potential investor situation and did not expect it would turn out to be a market.

For a while, we've been talking with some of our teams to consider the Middle East and Asian markets, instead of trying to compete in Western Europe, because we thought it might be easier to address these markets from Bulgaria then to address the US or European markets. Very few people, however, actually seriously tried to address these markets. And Sensika were one of the few teams that realized they couldn't do this remotely, and had to physically go and work from there. The same thing probably applies to every market—if you want to work on this market you have to be there and you have to make an effort from within, and not remotely. But then again, as in many other cases, the competition in the English-speaking world is much stronger than it is in the untapped markets of the Middle East. What is more, entering the US and Western European markets is often a matter of personal contacts, which we, coming from Eastern Europe, don't have. I believe it may be a bit easier to compete in the Middle East and North Africa markets, because although you still need contacts and a network, it is a level playing field where startups have the same chance to succeed, no matter which part of the world they come from.

Q. How do you see the role of investors and advisors in the development of Sensika?

A. Sensika are one of the teams that managed to take advantage, if not to the maximum but at least a big part of the resources that were made available to them by their investors and advisors. They have always stayed close and played a part in our community, have been asking for advice and discussed various matters with us and have managed to take advantage of the contacts to whom we have introduced them.

One such introduction was to UniCredit (a large European bank), which they managed to attract as an investor, when they were at the end of their resources and runway, so this is a good example of investors playing a role in the life of the startups.

But at the same time there was always a certain level of stubbornness with Sensika, as for example in the case of the prioritization of the scalability of their technology from the very beginning.

Q. What is the role of chance in the life of the startup?

A. It is not so much chance but your ability to take advantage of the chances that come your way. It is one thing to meet a partner or client, and it's a completely different story to be able to make something of that meeting. We have often said that as investors we can only open the door for our startups but they are the ones that have to do the work. It is the same with chances—they will come your way, but you need to take advantage of them.

Q. What would have been different if Sensika had started in the Silicon Valley?

A. They would have either died sooner or developed much faster and they would be a much bigger company now. The market there wouldn't let a company like that linger for four to five years before demonstrating some success. They are persistent and persevere through the difficulties, but speed of execution has never been a key characteristic of their company. And my understanding is that in the United States you either succeed fast or you fail—if you can't prove you are onto something within the first year, I don't think anyone will support you further.

Q. What was the role of the Bulgarian diaspora in the development of Sensika and the other companies?

Helping companies is to a large extent like investing in companies. Angel investors prefer to invest locally in projects they know and understand and in markets they know and understand. I believe that Bulgarians living abroad are more inclined to help Bulgarian startups purely on an emotional basis. It wouldn't matter if a person is from, let's say another Eastern European country, as long as he or she is not Bulgarian, the level of engagement with a Bulgarian will not be the same.

In the case of Sensika, they got in touch with a Bulgarian national living in Dubai and she helped them open many doors, alongside investing a small ticket in the company. So in their case the diaspora was very supportive and played a role in their launch in a unfamiliar market.

Q. What do you think about the team and what are the factors that drive them?

A. One thing that is interesting for us about Sensika is that they only got to substantial revenue five years after they started working on this project. This comes to show incredible perseverance and persistence. Almost every startup goes through highs and lows but having the guts to stay afloat for five years, at moments supporting the company with your own cash, is worth acknowledging.

In the case of Sensika, there is an accumulation of factors that play a role in driving the team forward. They are all entrepreneurs, and had been entrepreneurs before they started this project. So, for them it is not really an option to go back to paid jobs at this point in their lives, and the drive to succeed as entrepreneurs is a combination of pride and self-consciousness; each of the people on the team has at least one project behind them, and has accumulated enough experience that it is time for them to make something big, and failing to do that with Sensika would pose a challenge to starting something else afterwards. The combination of experience and background they have creates pressure to succeed.

11

Rentl.io

Company Profile

Product description: a software as a service tool for hotel and accommodation providers

Website: www.rentl.io

Founded: 2014

Company size (# FTEs): 0–10 | **10–20** | 20–50 | 50–100 | 100+

Funds raised (in Euros): **n.a.** | 0–100,000 | 100,000–500,000 | 500,000–1,000,000 | 1,000,000+

Investors: n.a.

Annual revenue (in Euros): n.a. | 0–100,000 | **100,000–1,000,000** | 1,000,000+

Interviewees: Marko Mišuli, Founder and CEO, Rentl.io; Ivan Brezak Brkan, Founder and CEO, Netokracija

V. Andonova et al., *Entrepreneurial Ecosystems in Unexpected Places*,
https://doi.org/10.1007/978-3-319-98219-9_11

Case Study

Rentl.io was founded in 2014 by a former finance professional turned entrepreneur. Its aim is to provide a full technology solution for small hotels and accommodation providers.

When funding became available to early-stage technology companies in the region in 2012 with the launch of Eleven and LAUNCHub, all the countries experienced a surge in entrepreneurial activities. Many wannabe entrepreneurs saw an opportunity to get support and resources for their plans and started building, or at least considering building a startup. Being a startup founder became a sort of fashion and some startup founders tried to become rock-stars, following the path laid by prominent entrepreneurs from Silicon Valley. And as these were the loudest and most visible, it inspired a whole entrepreneurship-as-a-lifestyle trend, where people were talking the talk of being startup founders, but not really walking the walk, as they were mostly focusing on attending startup events, winning pitch competitions and generally being in the spotlight.

Of the over 3000 applications we have seen throughout the three and a half years we were actively investing, we could clearly separate the applicants into three groups:

(i) people who were already working on a business and considered obtaining funding and investor support as a means to facilitate and accelerate their development;
(ii) professionals in established companies or corporations who had spotted a problem and believed they could address and solve it better by moving outside the organization and working on it as a separate business;
(iii) people who opportunistically applied for funding, usually with just an idea, without having spent too much time working on it, with the notion that if they obtained funding they would either get investor support and succeed, or fail on someone else's account, but in either case they were putting nothing at risk (except for the time to fill in the application and go through the selection process).

Not surprisingly, six out of the seven companies we have interviewed for this book fall either in the first or second category.

We have chosen to talk to Rentl.io, because they embody the exact opposite approach to the third type of startups, and we wanted to give the floor to the entrepreneurs you don't hear much about, because they are not the ones you'd meet at a conference or see on stage at a startup pitch competition; instead they are spending long hours working on their startup, speaking with clients and partners.

The company is a textbook example of a super-majority founder, with little to no experience in the domain of his startup, who is launching a company in a saturated and very competitive space (tourism), out of a small city in Croatia. When you put all these together, you usually get the recipe for a quick fail, either because of team friction or lack of product–market fit, for example. And yet, Rentl.io is still going strong, with significant traction in Croatia, able to support a team of nearly 20 people without any outside investment.

When talking to the founder, Marko Mišuli, and later with his colleague Ivan Brkan, we quickly came to realize the key differences in the way they are developing Rentl.io to most Balkan startups we have seen.

To begin with, Rentl.io's founder started by investing in building the prototype for his idea himself, before approaching investors or potential co-founders. Furthermore, he went door to door with his prototype, speaking to his future clients and understanding what their problems were and how best to help them solve them. These things are quite rare in our experience; more often than not, startup founders first try to find people who could fill in the gaps they have in talent and skills before spending too much time on an idea, and many times they would totally skip the customer development stage and dive into full-scale product development without having understood whether there is a problem to solve and how to solve it.

As Marko had a story to tell, his future clients were willing to spend time with him, listen and talk to him honestly about what they needed. This gave him another story to tell his future co-founder, and he managed to recruit a very experienced technical person to his startup early on. Finally, building that initial relationship with his clients helped build trust and facilitated referral marketing, where his first clients became ambassadors for the product to their neighbours and friends in the business.

Two more things stand out in Marko's case: his attitude to startup pitch competitions, which he has used to secure some cash (winning the prizes) for the business in the beginning rather than fundraising, and his approach to work–life balance for his team. As he explained during our conversation, he understands his team members have families and interests above and beyond Rentl.io, and he is happy to provide an environment where work and personal life can co-exist.

Lessons Learned

Startup life in the spotlight is tempting, and the startup founders who are doing entrepreneurship as a lifestyle, instead of actually investing themselves fully in their projects, are more vocal, so what you see in the media is not necessarily the essence of being an early-stage tech entrepreneur. The hard workers and dedicated founders are hard to meet, as they are busy with their projects, but they are more likely to be heard about later, when their projects succeed and grow.

Founder Interview

Q. What were you doing before Rentl.io? What made you choose to start your own startup? Tell us about the early days and how it all began.

A. I started Rentl.io in 2014, and in the previous eight years before that I was working in the finance industry, for a corporation in Zagreb. I was fed up with it at the end, as I didn't always share the same vision with the management of the company, but it was a good job and there was no compelling reason to quit.

I am originally from Zadar, a small town on the Adriatic coast and when my mother passed away, it changed my priorities and my entire vision of life. After spending all this time in Zagreb, I realized I missed Zadar. I had my family there, a small baby, and it seemed like a good idea to go back and start there. And after all, I wanted to prove that it is possible to start from any place and still be successful.

As a matter of fact, that was part of the motivation to start and get involved with entrepreneurship in the early days—I wanted to set an example. There is a lot of negativity and despair in Croatia, especially among young people, and a belief that it is not possible to start a business and to launch on a global level from there. This makes young people afraid to even try. So, I wanted to set an example of how it would be possible to start a technological company that targets the global market out of Croatia.

Q. How did you and your core people meet? Is the whole team still together and what does it take to keep a team together for so long? If core team members moved on, what were the reasons and how did you handle it?

A. I met my first team member, Ivan, at my previous job. We were not friends, but we had done work together and I knew that he had all the expertise needed to build the first version of the product. So, when I had a prototype and some positive feedback from the customer interviews, I approached him with an offer to join me on this project.

Being passionate about your product and showing people your vision is the key to attracting high quality people to the team, but on the other hand, as far as people are concerned, one thing I did not do then, and I never have done up to now, is to over-promise. And this is something I find very important: you should be honest with everyone—yourself, your colleagues, your users, and everyone involved. I never promised something I could not deliver and this gives people the opportunity to make their own choice.

Today, most team members in the company are from Zadar. Ivan, the first person to join, is not full-time in Zadar and works from Zagreb. And again, the key to keeping and motivating people is to not bullshit. There is a lot of startup wisdom on keeping and motivating the team, but it is not necessarily applicable for Zadar, or true anyway. People need to meet their required standard of living—they need a salary, and you cannot offset this by giving them stock options. It is true that at the end of the day every employee needs to feel part of the company to connect to it more personally, but you need to meet this need first.

Nowadays, the way we attract new people and keep our team motivated is to focus on the technology and the product and to keep the company ahead of the market. This makes us an attractive employer and helps in acquiring talent with good knowledge and expertise.

Q. What did it take to turn this idea into a business? What were the key challenges along the way? Did anyone help you with this?

A. I believe you should not be looking for investors before you have clearly defined your idea. There is a lot of startup bullshit going on, and people are hustling and pitching before they have done any work on their projects. With Rentl.io, I didn't even start looking for team members and co-founders before I developed a prototype of the product. Once I was ready I started going door to door at small hotels, showing my prototype around and asking for feedback, trying to understand the market better. And I only launched after I had accumulated sufficient knowledge and understanding of the market and enough positive feedback from those initial interviews.

As I mentioned earlier, the first person to join Rentl.io was an acquaintance of mine, Ivan. At the time he was working for a large digital agency. I showed him the prototype and pitched him the idea and persuaded him to quit his job and join Rentl.io full time. That was in May 2014, when we started working seriously on building the product.

From then until March 2015 we did all the work in order to launch a complete product. Ivan was doing both the front- and back-end of the website, while I was working on the marketing campaign, materials and other related issues. We didn't make any noise or do any marketing before the product was ready.

We didn't get any investment—not in the early days, and not now—we bootstrapped along the way. The only thing we had was our product—it needed to be stable, of high quality, and to bring value to our users. And it did. So our first users became our ambassadors. We only hired our first sales people in 2018; prior to that all our sales were simply based on word of mouth. Almost every house on the coast of Croatia has something to do with tourism, so we had a vast market to tap into. And also, people know each other, so if you solve one person's problem, he or she would quickly refer you to his or her neighbour, creating a sort of off-line viral marketing.

Funny enough, my quitting a well-paying job helped the initial sales a lot—people were curious to meet me and hear the story of someone quitting a good job in the capital to start a company in a small town on the coast. But this proved to our clients my commitment and dedication to what I was doing, on the one hand, but on the other hand again proves the lack of faith of people in Croatia in starting and being successful in a tech business, targeting the global market out of Croatia.

Q. Tell us about your first hire and your first fire. How do you approach this, and how did it change from the time you started to today?

A. (At the time of this interview) Rentl.io has 15 people on the team, no one has left and no one has been fired. I don't expect this to always be the case—as the team grows some people may decide to move on, we may need to part with others. And although I have not fired anyone so far, I believe this is part of the life of any company and also a personal challenge to myself, to help me grow.

We have one key advantage to most IT-related companies in Croatia. Many of those are outsourcing or development studios, working on projects for foreign clients. But when you only work on one small part of a project, you quickly get bored and don't connect with your job. You may end up working on three or four different projects in the same year, and your job is just writing code—you never get to see the final product, what it does and how it works, or get involved in discussing various aspects of this product that are not code. In Rentl.io, it's quite different: you participate in building the entire project and you see how the users engage with the product. And in the end, you get recognized for the work you do and the value you create. Ten years into the future you would be able to say, 'I have built that product, put it out on the market, worked with its users, and I have experience with every aspect of building and launching a product. This lets me do whatever I want in the future.' Participating in the entire process leaves you with the feeling that you leave a trace.

Now that the team is growing we have someone working on recruitment and selection, and selection is now becoming a very serious process. But we don't do any of the startup vanity team gimmicks—we do not have a ping-pong table in the office, we do not expect people to spend crazy hours at work—we respect personal time, family time.

When considering a new team member, we do look for culture fit, but our culture is just not the run-of-the-mill startup bullshit culture, where you get to drink beer at your desk or work from a tree. Still, we are looking for people that have an entrepreneurial spirit, no matter who they are and what position they will be in. Personality is a major factor, but we expect people to be more focused on their personal growth rather than hustling back and forth and having that startup buzz culture and attitude.

Q. You have not taken an investment in your company. What do you think about investors? How do they help, and how do they hinder a startup? How about mentors—where did you find yours and how did they help?

A. As I came from the finance and asset management industry, I have a very clear understanding of what investors should be looking for and should expect from an investment they make. It is striking to me how this startup investment market works—there are startups that would raise substantial amounts of money and still fail before selling even one dollar's worth of their product.

When I started Rentl.io I chose to go against to the prevailing practice; this is something I learned during my previous job: if you want to succeed you need to go against the current. When everyone is selling you should buy and vice versa. So we started by aiming to make revenue first, before ever considering raising money from investors. I believe this approach may make you more successful in the long run. When investors are involved, timing and development of a startup is totally screwed—you are expected to grow very quickly and, if possible, get to huge revenues early on, but it isn't possible to create a healthy company in just a few months—it takes a while to build up your team, your product and your business.

I get headaches from hearing startup pitches, when the majority of those are just plain PowerPoint presentations with nothing real behind them. In the early days we were somewhat active at startup events, but the only reason we did that is that we had a product, and when we were competing with idea-only startups it was easy to win, so for me using the prize moneys from these pitch competitions was a way to fund the development of the company.

But don't get me wrong—I am not against getting an investment in the company, but the time for that needs to come and it is not early on. I don't accept it is reasonable to get an investment before you have started—maybe other people or other markets may see this differently—it takes discipline to be able to take money early, and very few startups have that discipline.

Another thing is, that we keep hearing about 'smart money', but often they are not that smart. The majority of the VCs in this region do a numbers game—they invest in many startups and wait to see some of them grow. And there are many 'serial entrepreneurs' that have started and failed projects many times but are still around and trying to fundraise for their next project—it is not something that is healthy or makes sense.

And as far as advisors and mentors are concerned—there are many very smart people out there, and knowledge is so cheap and abundant—I do not have any personal mentors or advisors, but I read a lot, and this is enough of a base for me to form an opinion that I can apply in many situations.

Q. What was your darkest hour? Have you had a 'near death' experience in your startup development? What was the reason? How did you overcome it?

A. Every day is difficult when you are running a startup, and although some days are more difficult than others, there isn't one day or time in our development that stands out. It is an emotional rollercoaster, and if you like challenges it's OK—every day there are different decisions to take and different paths you can take. So if you are that kind of person, you don't see these as dark hours, but find the whole experience very satisfying.

Q. What comes next for Rentl.io? What would you do if/when you sell your business? If you weren't doing Rentl.io, what do you think you would be doing now?

A. The next thing for us, now that we have made some serious progress on the local level, would be to show up on the radar globally. There are some serious partnerships coming up, but it is still too early to tell.

However, I don't focus too much on the future; we have a global vision in mind, but our focus is on the day-to-day work, getting better every day. So, we keep our focus on the micro level, while keeping the big picture in sight.

Q. What is it that makes you get out of bed every morning? How is your product making the world a better place?

A. I want to create a product users love and this is what drives me and the team to go further every day—every piece of positive feedback we get, and not only from our users, but from everyone in the industry, is what keeps us going. And of course, I want to prove that it is possible to start a global business from Croatia and hopefully inspire every young person in the country to go into entrepreneurship.

Q. If there is one thing you know now, but wish you had known at the start, what would it be?

A. It is key to define what success is; for me, success may be completely different from what success is for someone else. And this is what I tell people—Rentl.io is very small, so if you define success as having huge revenues it is not successful. But as a person, I am super successful—I have this job that I love, I have time to spend with my family, and I get to live in the best city in the world. That is what success is for me.

Q. What do you think would be different if you launched Rentl.io out of Silicon Valley, for example, instead of Croatia?

A. We were very happy to start in Croatia. Early on we had the opportunity to join a globally recognized accelerator program in the USA, but we decided to pass and work from here, and to this day, I still believe Croatia was the best place to start. Maybe in some markets we would have easier and better access to investors, and there are definitely markets where legal and administrative matters would be much better organized, but Croatia is a great market for tourism and for this particular project it was the most intuitive and natural place to start.

Advisor Interview

Q. Rentl.io is doing things differently from other startups. What do you think about that?

A. I met Marko some time ago and the thing that stood out, even at the beginning, was that it was an unusual story in several aspects. The first was that they were doing something for which there was a market in Croatia—tourism. Tourism makes up about 20 per cent of Croatia's GDP, you have customers here, you are creating a product that you can promote locally and you can actually make money out of it. And they are making money locally in Croatia. This is really different from all startups focused on the US market.

The second thing was that Marko is not an engineer, nor is he the college kid trying entrepreneurship for the first time. He is a finance person, but in the sense of a classic finance investor—he is more of a business person who has the right mindset. He already had a career, a successful one at that, and that decided 'Okay. I now want to have a company'. So, with Marko it was not like with a lot of stories where you have people out of college or high school, thinking 'yes, let's start a company!', while having no idea what to do.

The third thing was that he was based in Zadar, which is a small city on the seaside, and it's basically very unorthodox. Most tech companies are based in Zagreb, maybe Split or Osijek. Having a tech startup in Zadar was very unusual. So that was what was different about him.

Also, he was a person with the right mindset. He had enough life experience to know what was right, but also a good head on his shoulders in terms of how to deal with people and investors. He wasn't an inexperienced kid, but most of all, he was an honest founder.

So what he did was going against common norms and he was succeeding in it.

There are couple of other startups they are doing kind of the same thing, but Rentl.io was one of the first who was actually making a product, while the others were making agencies. He is actually making a product that is well designed, he got a really good Croatian CTO, which made me believe that he was really serious about this company.

Q. What do you think they are doing better than anyone else?

A. One thing is, he is definitely not subscribing to the startup hype both locally and internationally. As I said earlier, one thing they do differently is that they are working in an industry that makes sense in Croatia, that you can test, that you can develop and in which you can talk to customers. And again, Croatian landlords are really fickle. If you can sell a product to them, you are actually going to have a much easier sell to others. We, as the media, promote a lot of tourism-related products for big companies and one thing that I hear from a lot of them is that selling to people locally is really hard. When the tourist season starts you cannot sell anything.

In this specific case what he actually did was he went out in the field. He talked to all of the customers, he sold door to door. Selling to customers in terms of actually knocking on people's doors is something that most tech companies in the Balkans don't do. They are actually scared, don't want to be proven wrong, they don't want to do the wrong thing, so they don't want to talk to anyone. They are antisocial. They want to code something and be perfect from the beginning.

Another thing that is different about them is that they are experienced people. Both Marko and his co-founder, as well as everyone else on the team, are experienced. These are not kids playing with frameworks and business models. He actually knows how to create a business plan, but not in the sense of writing a 20-page document. He actually understands the industry and he knows that the tourism industry needs a product like that. And their app is really amazing. It is actually built really well. It is like going on Product Hunt and you see something really awesome and you wish you were doing something like that so you can use that exact product. I have this experience with a couple of apps, and in his case, I feel like being a landlord, so I can use that product.

The other thing about Marko specifically is that you have a person who understands his product very well and on the other hand is really humble. He knows what he knows, he knows how to make the case for it and how to talk about it, but on the other hand he is not the classic Balkan startup founder who's too fond of his own stuff and believes that he has all the knowledge in the world because he has passed through one accelerator. I hate that sort of founder and we have too many of those in the region. When you talk to Marko you know that he is actually listening to you. He wants to hear and learn from other people.

Q. Croatia is a fantastic market for a tourism tech company and you'd expect to have a myriad of those, but there are not all that many. Why do you think that is?

The people who want to make money in tourism will find a way to make money in tourism that's easier than making a tech company. I just think that for a lot of people having an apartment to rent makes far more money short term than having a tech company. And these people will not be into tech.

On the other hand, the ones who want to create a tech product won't, because it is Croatia, and it is a small market and everything that we want to create is for a bigger market. They are going to go into other industries besides tourism. Basically if you are interested in tourism you are going for tourism, not tech, and if you are going into tech, you are not going to be interested in tourism. That's why you have most of the tech companies based in Zagreb and Split.

Q. How do you see yourself, as an advisor, and the other people who played a role in the development of his company?

Well, one of the things that I discovered was that Marko actually turned out to be a product guy. He had never worked in tourism, but he had a lot of professional experience and a vision to build a product in this industry. He can easily tell who is talking to him for the sake of talking, so he won't take advice from those people and I think that, unlike many people, he listens not just to humour you, but is actively seeking your opinion, which he would then process and apply to his project.

Q. How do other startups perceive him and Rentl.io?

The true startups, the true tech companies, the true product companies, which are few in Croatia at this point, like him and he likes them. The startup lifestyle guys don't even see him or them. There is a big divide. Before, the startup founders all kind of mixed together and it was one big group. Now it's not. There is no startup community, in my opinion, people don't know each other. On the one hand you have people who are very successful and on the other hand you have the new startups. And a lot of those people are not talking to each other. So I would say that a lot of these big tech companies in Croatia have no opinion on Rentl.io. But the companies that are also doing cool products respect him and he respects them. It is a small, but really good group of product companies in Croatia.

Q. When do you think will it make sense for Rentl.io to go and fundraise and what will it take for them to be investment ready?

I follow the tourism tech sector a lot; it is something that interests me and I think there is a lot of wasted potential in Croatia. I think that Croatia could have a lot of tourism tech startups. It was probably two years ago when I wrote in an article that the huge Croatian holding companies with a lot of money should invest in tech startups.

One thing Rentl.io needs is that in order to grow (not to get an investment, but to grow), they need to build a marketing team. They have a product that's ready and what they need now is a growth strategy. And then, when they find a strategy that works for them and they actually understand how to grow, then they will be ready to attract investment and use the money to grow further.

The other thing is that Marko needs to transition from founder to CEO. And I know that he knows that and he is thinking about taking this step. He needs to move on from taking care of every single problem in the company to actually focus on the things that will make the company grow. And this is one of the things that the founders in Croatia, including myself, struggle with. We are doing things we shouldn't be doing instead of doing things we need to do. Or we are just content to stick with the things we've previously done. I think that Marko needs to make this transition, focus on what he is best at, expand the growth and marketing team and then he will be ready for an investment.

And it needs to be smart money, someone who understands the potential of Rentl.io, and then there are very interesting things that could happen. I really think that is going to be an interesting story. They are going to be a tech company that is going to conquer the European or the Asian market in terms of what are they doing.

Q. Imagine Rentl.io had started in another place, Silicon Valley for example. What do you think would have happened?

I think that he would be based there, stay for a bit and then move to another tech community in the USA or maybe Canada and start a really good profitable product company. He would understand very quickly that the people in Silicon Valley are not going to help him and he would move away and do things his own way.

12

Telerik

Company Profile

Product description: software tools for web, mobile, desktop application development, tools and subscription services for cross-platform application development

Website: www.telerik.com

Founded: 2014

Company size (# FTEs): 0–10 | 10–20 | 20–50 | 50–100 | **100+**

Funds raised (in Euros): n.a. | 0–100,000 | 100,000–500,000 | 500,000–1,000,000 | **1,000,000+**

Investors: Summit Partners

Annual revenue (in Euros): n.a. | 0–100,000 | 100,000–1,000,000 | **1,000,000+**

Interviewees: Svetlozar Georgiev, Founder, Telerik

V. Andonova et al., *Entrepreneurial Ecosystems in Unexpected Places*,
https://doi.org/10.1007/978-3-319-98219-9_12

Case Study

Telerik is one of the success stories in Bulgaria, after building a global business and selling it to Progress Software for over US$250 million in 2016. Because there are numerous things to learn from the Telerik story, as they have walked the whole way from start, through growth, to exit, we asked the founder to focus on the one aspect that he believes is the most important and relevant for other early-stage companies, so he chose to talk about the 'whole product' concept. We will not be analysing his interview here, because the conclusions and learning outcomes from it are very straightforward and clear.

Founder Interview

Q. Tell us the story of how you started with Telerik.

A. It all started by chance, as with many things in the genesis of Telerik. The question is always how you use and develop this chance.

We founded the company by chance, because we suddenly quit our former company. With the other founders of Telerik, we were working in another company, a branch of a small UK company in Bulgaria. There were some issues, as the company was run remotely from London via ICQ and the style of management was not the most suitable, to put it mildly. But we wanted things to go well, as it was one of our first jobs and there was a strong sense of loyalty, so we made a rather constructive list of issues and solutions we proposed. So one of the team sent the email and as a result got fired for his audacity in doing such a thing. We had no other choice but to quit as a team, and we found ourselves without jobs, which was not a big issue as we were still living with our parents.

But we thought, if we think we are so smart, why not try to do something ourselves. The first thing we did was run projects—websites, systems, whatever came our way. We probably did four or five of these.

The story gets even funnier. One of the founders, the only one who was not from Sofia, was looking for a place to rent, and another founder's mother had a real estate business, so she agreed to find him a place without commission, in exchange for a three-page website. So, as the . NET platform of Microsoft was just out, while he was doing the website he was geeking out and testing this and that and made a component for content management, a primitive HTML editor that was pretty easy to do with the new platform. He showed it to us and we liked it a lot. We figured there were a few companies that were doing a similar thing and selling it, so we decided to package it and try to sell it as a product. We set a firm deadline: if we don't sell anything within a month we need to do something else. During this month we sold a few licenses, even a professional one that was US$349 (as compared to the other version that was US$49). We kept pushing and more and more clients came in. It was a very small revenue, but while we had the opportunity to take some projects to work on, we decided to focus on selling a product instead of providing services.

The curve started going upwards, but anyway we had a very low cost base at the time, as we were staying in a room in my father's office for free, we got the furniture as a gift from my wife's company that was renovating its office, we were sharing an internet connection with four other companies, so we were doing a classic bit of hustling wherever we could. Anyhow, the decision to go for a product and not a service business was taken, and this is more or less the genesis of Telerik.

From the very early days, even when we were just the four founders with an employee or two, we pushed heavily on support. When a client came to us, until his product was finished, we did not let go. It sounds logical, but not everybody is doing it, as it takes a big investment.

Another thing we did in this aspect was the organization of the work in the company. Everyone in the company was doing support and answering tickets, all the way up to the CEO (however funny it is to have a CEO of a ten-person company.) The developers were developing half of the time and closing tickets the other half. However difficult this was to impose, as developers want to be creative and to write code and not deal with clients, we were very keen on having it this way and having everyone face what our clients were facing, to hear their problems first hand and then address them.

This was a great help for product development, as you have very short feedback loop and you can see what is happening immediately.

Furthermore, if you release something stupid, whether it's a bug or a release that is not well thought out, and you get five hundred tickets the next day and you have to close them all, the problem slaps you in the face, but there is no one else to blame. Otherwise, if you have a support team and a development team there is always a blame game and each team is pointing at the other should any such problem arise. Our approach blurred the boundaries between the teams and created quite a strong feeling of responsibility.

Q. Did this help handle team dynamics?

A. Yes, it did, but another thing we did that also helped a lot was that we grouped the teams not by roles, but by product. While we had distributed the support throughout the company, when we grew a bit and increased our product portfolio, we had grouped the teams by roles: developers, QA, etc. But at some point this structure started creating friction, allowed for blame games and was unhelpful for alignment of goals and timing. So, what we did was to group teams around products—each team was composed of the right people to bring this product to market, with minor exceptions—for example teams had no accountants, but had everything else.

So, each team knew what their product was, who was their client, when was the release date, and all these people were working in the same room, spending their entire time together. And as the release date approached, they knew what each of them had to do, as it would affect everyone on that team. And each team member knew everything about the product, as they were spending all this time together.

So, by grouping people this way, it worked miracles. Again, it was not something we read in a book, but we did it on a gut feeling and we nailed it.

Q. Was it difficult to take the decision to do something against the standard practice at the time?

A. I think it was quite easy, as we were rather irresponsible at the time; we had a sort of cowboy style of management. We were all relatively young and inexperienced, so we had neither the prejudices, nor the knowledge and routine to know which way was the right or the orthodox one. We would feel something was right and we were quick to give it a try. But we were trying to be as open as possible with the teams, so if something did not work we could address it quickly, but also to share the responsibility to some extent, because it is very easy to make a mistake and we did not want to leave the impression that someone on the top floor came up with something and it did not work. Instead we would engage the teams and tell them that we were going to try this thing and see if it works or not. So, more or less the lean and agile way, without knowing what lean or agile means.

Q. How did you approach building competitive advantage and speed at the beginning of Telerik?

A. Our core product is just a piece of software. In certain aspects it is better than competitors', in others not as good, but in this the competition is never-ending. One release you are better, the next the competition catches up, then you innovate on something. The area where we innovated, through gut feeling more than anything else, as startup literature available at the time wasn't so easily accessible, was the so-called 'whole product'. A client does not simply buy a piece of software, he or she wants to solve some problem, and to solve this problem you need to provide a number of other things around it.

One thing that is critical is that first you should present the end result of what the product does. And while our competitors were focusing on the fact that they have a certain number of features, our focus was on what the software one can build with our product could look like. So, we innovated quite a lot on examples and demos; we focused heavily on solving complex cases, very well designed, and when clients saw them they would think 'I want that!' If you swapped our product with other products underneath, you'd get to the same result, but our rivals were presenting something ugly, the so called developers' design, while we were creating a desire in the client: 'I want that. I like it.'

We did this very early; we realized that everyone else was presenting their product in a rather unattractive way and maybe our background played a role here; my first and only job before Telerik was in marketing. I was working for an international marketing agency, where I quickly moved up the ladder to become account director. For me this was very helpful, as on the one hand I felt I had a very good theoretical foundation from the American University, and on the other hand there were a million things I thought I knew, but in reality I didn't. Until you face the client and the reality, it is hard to say you know what you are doing at all. So, I had some affinity in me with respect to how we could present something well, and appeal not only on a functional and technical level, as in, I have these three features, this is my API and these are my technical parameters; this is something everyone was doing. But at the end of the day, our clients were creating end-products, and we presented to them what their final result could be. Our message was 'your product can be this good, it can have this appeal, this functionality.'

So, this is one of the key aspects: it was not the core product, but the packaging, as part of the whole product, that we focused on in our presentations. And while everyone else, including the big players, was positioning their products in a technical manner, it felt natural to us that this is the way we should market it. It is just like any other product: you can have the same thing inside, but you will make your choice depending on the package. You can say the same thing about smartphones: their functionality is more or less the same, but the brand weight of one is quite different from the other. Again, these are not functional characteristics, but ones that play a role on an emotional level, and that, at the same time, present what the product does in an excellent way.

The other thing we did, and maybe it was even more important, as at some point our competitors realized what we were doing and followed suit (it is a never ending chase there), was support. Again, this is not core product, but it is part of the 'whole product'—the shell, which you can live without, but which is once more critically important to have, so as to be able to provide value. The value here is for the clients to be able to do their job. They don't need the binary (the executable), they can get that for free from Microsoft, but when they have a problem, there is no one to help them out at Microsoft in the middle of the night. What we did, and it again happened naturally as the team took the problems of our clients to heart, was to help the clients in any way we could. For example, if a client had a problem that was unrelated to our product but was simply a blockage for them to finish their job, we would help. We would write the code, the client would give us a problem and we would write the final product. Some clients were even abusing that: they were throwing us an assignment and expected us to finish it for them, and we did. (Later on we had to address that, as it was wasting a lot of our time.) But when you help your clients above and beyond what they expect, they become your best evangelist. This is why we had unusually high level of referral sales, because people were just referring us to others, or when changing jobs they would recommend our tool in their new company, because they felt us not just as a vendor of the binary, but as someone who was there to help them do their jobs. This is something that came up very often when asking our clients why they chose us. In their view our product was good, but the competitors' products were very similar or the same, and Microsoft was providing these for free. And this is something that happens more and more often; software is commoditizing and becomes free, because other business models are supporting it, as is the case with Google, which would give everything away free, because it makes its revenue from advertising. So, our clients were not paying us for the core product, but for the attitude and the support, because they know that with us they can get the job done.

Q. You've had an investor in your company. What do you think about investors? How do they help, and how do they hinder a startup?

A. I think the role of investors is critically important; for us it played a pivotal role in many aspects. Overall, our investors helped us become a real company. As I mentioned earlier, before that it was a cowboy story—yes, we were all friends working together, we had a great culture, but we were not at a global level on management, governance, vision, or mindset, if you like. And I am certain that if we had kept on going it alone, we would have either hit a plateau at some point, or fallen apart. It may be a cliché that the investors take you to the next level, but that is exactly what happened for us.

In one aspect, getting an investment helped us psychologically, both us as the founders and the management, and everyone else on the team. For each of us this was the ultimate attestation that yes, we may not be working from a fancy office, but we are still a world-class company. Despite the fact that we were making good revenues at the time, we didn't feel like a real company yet—those were just some numbers. When we received investment help, it gave us a very strong validation that not only gave us confidence, but also charged us with energy for the next sprint. We were already six years down the road with Telerik at the time of their investment, and there was fatigue and some friction, some people were pre-burn-out, but when the investment happened, it was just like when you are swimming and almost out of breath, and then you break the surface, take a gulp of air and dive again to continue swimming. And this happened for everyone in the company; we virtually restarted the company.

The other thing that happened was the push to become a real company from the perspective of governance, processes. And this is not something the investors give you; they just ask you to do things, and in order to provide those things, you have to learn the hard way how to do them. It was very hard in the beginning, as we were not on the level that we should have been, and whenever our investors asked for something for the first time we had no idea how to do it. But gradually it became a cycle and started becoming easier.

Another thing investors help with is that there is another person in the room you need to convince about your plans. In this, our investors sometimes did well, other times not so much, because we were quite a successful company, and then you always get some slack and they let us do some stupid things. So, having an investor is no guarantee you won't take bad decisions, but neither is it the full freedom to take any choice you like just by yourself.

Finally, being a Summit portfolio company is a completely different story from being just some company from Bulgaria. When we speak to investors, partners or clients, they perceive us in a completely different way. And this is something you cannot buy with money or something you can ignore. This helped us hire a very senior executive team (CFO, CMO, etc.), all people that have been with one or more public companies. They wouldn't have paid any attention to us if Summit were not on board.

So, again, it is critically important, but an investor is a very broad term, and there are the very early stage investors that help you pick up when you are just starting. This is something we did without an investment, but getting a Series A investor is a must to get to the next level as a company, unless you are coming from a very strong school, you have a natural network and these are all resources you have available.

What is more, even within the same investment programme, you can deal with different teams and it may be a completely different experience; when working with investors you don't work with the brand, you work with people. Luckily for us, the people on our team are great and we have had and continue to have an excellent relationship.

Q. What was is it that made you get out of bed every morning, when you were already a successful company?

A. At some point the responsibility becomes so big, that you have no option but to carry on. We have had very hard moments and periods of time when going to work feels like a burden; a large company, with many people to deal with, an investor to handle, the risk of what may happen the next day if, let's say, Microsoft did something. We have even had our share of patent trolls to deal with, which is always tough. But it is a big responsibility and there is no way out. You have started it and you have to keep going, no matter what.

What gave us strength here is, that the core team stood by each other in these moments. Any one of us alone would have failed, but as a team we are very strong. Also, the second level core team, our initial employees and shareholders are great and we are still working together, so we always know there is a team to rely on.

Finally, we have always had a next goal to achieve and it was never some global distant vision; we were always aiming for the next hill, not the top of the mountain. So, our goals were always visible and achievable, rather than something vague like 'we want to be a billion dollar company'.

Q. Did the diaspora help and how?

A. Yes, on some micro level it did. In some places we set up our offices through friends from Bulgaria we knew in that country or city. Because usually the first person to plant the flag in a new territory does not need to be a domain-specific expert or some super senior business development person—these you can hire—but you need someone with whom you have a great level of trust. You need someone who is like a blurred copy of yourself in order to make it work, and for this person to be able to convey the company culture in this new market.

So in some of the core markets for us, like the UK, USA and Australia, our offices were kick-started by Bulgarians, and it wouldn't have happened without them.

Q. What do you think would be different if you had launched Telerik out of Silicon Valley, for example, instead of Bulgaria?

A. I think the spirit of unity and responsibility would not have been as strong in the Silicon Valley; when something cooler pops up in Silicon Valley, it is easy to lose some of your key people. There, the level of loyalty and affiliation with the cause is much weaker than here, and people are always looking for the next big thing or the bigger paycheck. Here, people would stick closer to the cause, and although our success brought along a certain level of perks, I don't think this was the major factor, as we have gone through some very tough times with these people and people have stayed with us.

Q. If there is one thing you know now, but wish you had known at the start, what would it be?

A. We were taking things very seriously and we were very nervous about everything. Although we had made a great deal of cowboy decisions and bold moves, there were things that we were afraid of, like firing a person we knew we should get rid of, or what our competitors would do next. Overall, the level of paranoia was very high, and we had a lot of anxiety that we could have spared ourselves.

Part III

Conclusions

13

Concluding Thoughts and Lessons Learned from the Birth of Entrepreneurial Ecosystems in Unlikely Places

Why Did We Write This Book?

There are several reasons that triggered our ambition to write this book. Firstly, we wanted to support and encourage the exciting entrepreneurial dynamics in the Balkans, celebrating the inspiring story of the birth of entrepreneurship in the region and telling as many stories of entrepreneurs, startups and entrepreneurship community members across the countries on the Balkans as we could. Secondly, we aimed at entangling these experiences into a big-picture look of the developments that have led to the current state of entrepreneurship in the region. Eventually, we aimed at extracting lessons learned that could be of value to decision makers, scholars and members of entrepreneurial communities from other parts of the world where entrepreneurship is seen as a positive force for economic and social change but is in its early stages.

Specifically, the trigger for this project was a conversation with Bill Aulet, Managing Director of the Martin Trust Center for MIT Entrepreneurship and Professor of the Practice at the MIT Sloan School. Veneta was using Bill's book 'Disciplined Entrepreneurship' for her Entrepreneurship class at the American University in Bulgaria where the two of us were teaching. Along with teaching our regular classes, we were experimenting

V. Andonova et al., *Entrepreneurial Ecosystems in Unexpected Places*,
https://doi.org/10.1007/978-3-319-98219-9_13

with numerous non-curricular formats and initiatives that aimed to encourage entrepreneurial and strategic thinking among our bright and creative students. As part of one of them, we held a virtual talk with Bill that exposed our students as well as the startup community in one of the first co-working spaces in Sofia to Bill's philosophy about the importance of taking a systematic and disciplined approach in every entrepreneurial project. We held some preparatory discussions with Bill to give him a more detailed background on the startup dynamics in our region, and during one of them we asked him what he thought was important for young ecosystems like the ones in the region. His immediate response was: *celebrate local success stories.*

We reflected on this insight and soon it turned into a powerful message for us. It became the spark for our idea to initiate a project of celebrating the regional entrepreneurial stories. While looking at statistics and indicators that reflected some of the elements of regional entrepreneurial ecosystems, we started to collect and systematize the stories, and personal philosophies of entrepreneurs and other members of the entrepreneurial communities in the Balkans. We began realizing that widely used metrics were of little value in terms of supporting and nurturing the ongoing developments in the regional entrepreneurial ecosystems. They were distant and coarse proxies that did not reflect the energy, hopes and passions of the Balkan entrepreneurs we knew. For the most part, these proxies were signalling pessimism instead of inspiring further desire for change. We felt that the story of Balkan entrepreneuring had to be told differently. For the sake of research rigour and comparability, in the first part of this book, we use well-established proxies and measures of entrepreneurship, but we also introduce different reference groups as benchmarks—the Baltic states and Central European countries instead of the Silicon Valley, the UK and Germany. More importantly, the second part of this book represents a collection of cases, interviews and lessons learnt that illustrate the energy and accumulated knowledge in the Balkan entrepreneurial ecosystem in ways well-accepted indicators can't. This is also a celebration of the remarkable advancement and even certain sophistication that the Balkan entrepreneurial ecosystems have reached in a very short period of time. This is a message of optimism for all but most of all for the young people in this part of the world.

The importance of celebrating local entrepreneurial stories and heroes made a lot of sense in the context of our efforts to nurture strategic thinking, entrepreneurial curiosity and innovation among our students and the early-stage ventures we mentored. Many of them would frequently refer to the stories of the superstar entrepreneurs of our time such as Mark Zuckerberg (Facebook), Larry Page and Sergey Brin (Google), Travis Kalanick and Garrett Camp (UBER), Brian Chesky and Joe Gebbia (Airbnb) and so on. Over time, however, we began noticing that there was nothing that energized students and young entrepreneurs more than meeting local experienced entrepreneurs and startup founders who shared their stories about pursuing dreams and testing their smarts in the regional socio-economic realities that were markedly different from the ones in Silicon Valley, Berlin or London. We kept organizing and participating into more and more such meetings with entrepreneurs from the region, and the energy that they consistently generated fuelled our own dream of doing something that celebrated the region's heroes in a way that would inspire more and more young people to become such.

Our Own Story and Lessons Learned

In the years when the two accelerator venture funds Eleven and LAUNCHub catalysed a lot of buzz and growth in the regional ecosystems, we sought our own opportunities to contribute to this dynamic. As academics and investors we saw several natural roles for us. The first was to help nurture the next generations of entrepreneurs who will join the ecosystem bringing new ideas, more advanced skills and entrepreneurial mindset. The second was to contribute with knowledge and understanding of the specifics of local entrepreneurial processes. This second role was more complicated and crucially important because in the realities of the fast growth of the ecosystem, most members of the entrepreneurial community were focused on fast-spinning everyday dynamics with little time to systematically observe and think deeply about the evolving situation and its implications.

As we started to explore and apply research frameworks that would be of value to the ecosystem, we began looking at comparisons with more advanced entrepreneurial latitudes. We worked together with key local stakeholders and obtained access to data on startup profiles, and conducted many in-depth interviews with ecosystem members to study regional patterns. As we were performing different analyses and comparisons, we started realizing that some benchmarks made little sense in the local context. Some of the indicators that were expected to signal success in advanced ecosystems were not necessarily all-positive in our case. In other cases, indicators that were considered as signals for concern elsewhere were not real barriers for local entrepreneurs. This made it hard to perform direct comparisons and benchmark against the mature ecosystems. A complementary approach to understand Balkan entrepreneurial ecosystems was needed in order to capture the specifics in the context. Systematic case studies and interviews were much needed.

While working with different members of the ecosystem and observing the developments, we also began noticing something else. There were very different voices across the ecosystem with different claims and points of view. Some were very excited about the rise of the number of startups, the growing visibility of accelerators, the exits and the dense calendar of events and meetups. They were optimistic and tended to emphasize that growth in activity was faster than that of the leading ecosystems in the world. Since we were on the Balkans where pessimistic outlook is more common, we also had the negative voices that undervalued successes, and emphasized facts and comparisons that sounded depressing.

Seeking to reconcile the observations and evidence we had collected about the regional ecosystems, and their comparison with more advanced ecosystems elsewhere led us to an important realization: understanding the real achievements and real challenges of the entrepreneurial ecosystem in the region was only possible on the basis of local specificities, and in the context of the broader historical, and current socio-economic dynamics. While benchmarks inspired by mature ecosystems were important as guidelines, there is a real need of realizing that a young entrepreneurial environment is different and evolves through steps that are conditioned by the local strengths and challenges, as well as the way entrepreneurs and other decision makers address these. The processes rather than their outcomes are hard to measure with existing indicators and datasets, but they seemed to be logical and important for understanding the developments in the region.

The following pages summarize some of the lessons we learnt about the specifics of our young ecosystem. We look at the elements of the ecosystem introduced earlier in this book and share insights about what seems to have mattered for entrepreneurship in these latitudes. Figure 13.1 summarizes the most common explanations for the rise of entrepreneurship on the Balkans based on our interviews. Given that the main point here is about the uniqueness of the path each ecosystem follows and the need to analyse it in its own context rather than religiously benchmark against the outcomes achieved by best, we cannot suggest that these observations will automatically apply to other young entrepreneurial settings. We hope, however, that they will provide some leads and direction for targeted analysis and experimentation, as well as for understanding

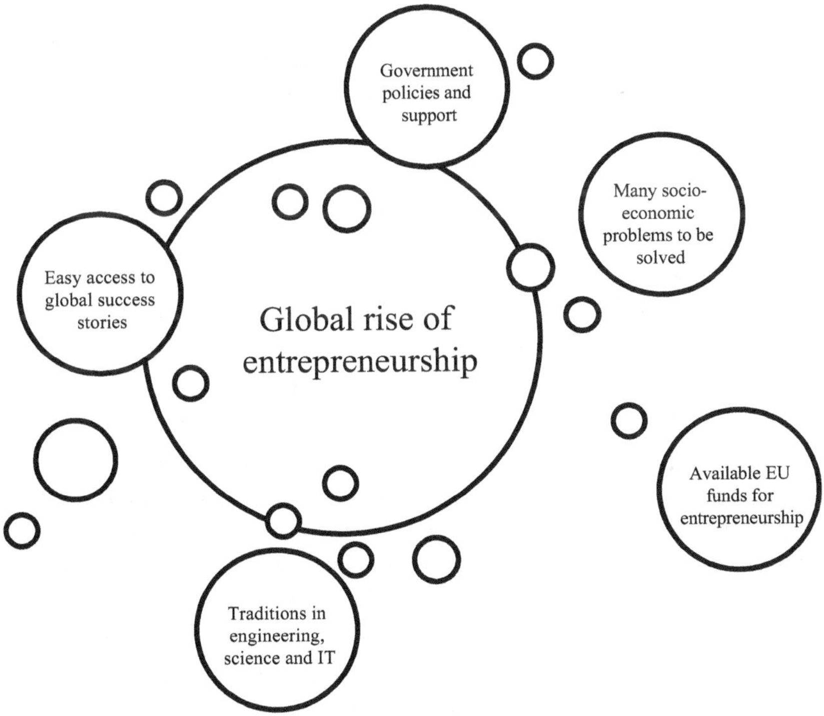

Fig. 13.1 Visual summary of the factors identified by regional ecosystem leaders as contributing to the growth of entrepreneurship in the region

how important early recognition and support is when aggregate outcome measures fail to reflect the energy and hard work of young entrepreneurial communities.

What Mattered?

As noted several times throughout this book, one of the important milestones for startup dynamics in the region is the availability of financing channelled through two large early-stage accelerators and a few other funds for later-stage investment. There is no question that availability of financing of this scale was an important factor for the success in the region. What is worth noting, however, is that it was pointless to judge the success of this financing through the prism of standard success rates and ROI measures applicable to accelerator funds in more mature places. Much of the available funds ended up supporting startup teams whose ideas did not take off at all or such that achieved only partial success. That, however, is only part of the story of what actually happened. Many of the entrepreneurs who failed did not give up and stayed in the community and moved to new ventures and new roles; then there were the ones who leveraged their contacts and formed new teams to launch new ventures. Some reflected on their lessons learned and began offering support services for other entrepreneurs; others became mentors and helped others overcome typical challenges. In that sense, funding failures gave birth to further expansions, growing sophistication of the entrepreneurial community and nurturing of a true entrepreneurial culture that permeated several societal groups, triggering new attitude towards entrepreneurship, which in the eyes of some observers can be called a cultural change. These important effects are hard to measure with standard indicators but were essential for the growth of young Balkan ecosystems.

The availability of financial resources for early-stage startup projects had another important socio-cultural effect. It stimulated many who would otherwise not sacrifice everything to become entrepreneurs to 'give it a shot'. This included many young professionals with successful careers at large IT companies or multinationals. This, in essence, accelerated the formation of an actual entrepreneurial community, which became a visible and active segment in the society, as it engaged some of the best-educated and active young people with a well-defined civil culture. This was an important side effect of the availability of significant early-stage financing as it legitimized entrepreneurship among leading social groups.

As noted earlier in the book, there are a lot of debates about the role of government and public institutions in stimulating entrepreneurship. Many believe that the best Balkan governments can do is to stay away from the community. A more careful look at the dynamics on the Balkans reveals that the public sector played an essential role in the success of entrepreneurial processes and that not getting in the way was only one possibility but probably not the best one. Public officials across the region have made decisions and encouraged policies that have helped fuel growth, including through the design of innovative financing mechanisms that were novel for government structures and programmes for the entire continent of Europe. Consolidating and directing investments towards infrastructure that supports entrepreneurial and innovation processes is also a factor. Governments are also the decision makers when it comes to longer-term policies and investments, for example, in education. Despite the fact that some of the encouraging developments with entrepreneurship curricula at schools were initiated by non-government organizations, without the involvement of public institutions, they would not be properly executed and supported with relevant policies and investments.

One of the findings related to the role of governments in the development of the Balkan entrepreneurial ecosystems is that a few well-trained and visionary public officials were instrumental for the most positive developments in the domain. As interviews with current and former public officials revealed, some countries in the region were 'lucky' to have young bright professionals as public officials at times when important decisions were made about financing mechanisms, infrastructure investments and business policies. Thanks to their top quality foreign education and previous experience with entrepreneurship, they acted like institutional entrepreneurs and made decisions that were essential for the ecosystem development. To reflect the full reality, we have to recognize that these lucky circumstances contrast with many other cases when government structures were not effective and were slowing down rather than facilitating positive change. In that sense, the valuable lesson learned is that governments can play a role and that there is important role that smart, entrepreneurial politicians can play in the public sector even if they are in office for just a few years. A second lesson learned is that encouraging entrepreneurship requires creative and innovative approaches that sometimes break conventional public sector thinking and models. The financing mechanisms that were designed and experimentally applied in the region produced tremendous effects and served as example for solutions applied to other parts of Europe. In that sense, entrepreneurial ecosystems need some innovators as public sector pioneers as well, and there is much impact they can have if they can be in the right political roles at the right moment.

As noted throughout the book, one of the most important advantages of the region in terms of potential for entrepreneurship is the availability of many well-educated and skilful professionals especially in the IT area. Advanced engineering and IT skills are at the heart of most of the success stories across the different countries in the region. At the same time, however, one of the gaps that remain unfilled over the last years is the limited business and go-to-market capacity in the entrepreneurial community. While the region has produced a fair amount of success stories and enjoys vibrant startup dynamics despite the lack of business sophistication, it is an area that needs improvement. Given that the traditional educational sector is slow in adopting reforms to address this issue, this is another domain for creative and entrepreneurial solutions. One possible approach that we have promoted in the last years is to integrate hands-on business learning for university students with support for early-stage ventures, where bachelor students act as business consultants supporting startups in their marketing and internationalization strategy domains. This model worked brilliantly and produced benefits for all participants. Students were excited to learn by doing and to apply what they have learned in the classroom in a real-world setting. They were also getting a taste of the world of entrepreneurship, were frequently 'infected' with the startup bug and later pursued their own projects or joined existing entrepreneurial ventures after graduation. Entrepreneurs got engaged with young talent they later recruited and saw the value propositions of their ventures through the eyes and smarts of the younger generations of business-trained graduates.

Ideally, such programmes that offer a creative compensation of the weaker business capabilities of entrepreneurs should be part of the binding agreement between investors and early-stage ventures in the region and should not be optional. This will ensure that even if a startup does not make it, its founding team will have more business preparedness and go-to-market skills to learn from failure and succeed in the future, while university students will get first-hand exposure to entrepreneurship.

Another approach to nurturing stronger market culture is to expand the format of non-degree training solutions that have risen across the region. The model of the discussed earlier Telerik Academy has grown as an entrepreneurial solution to the fact that the traditional education system does not train the skills that are demanded. In that sense, visionary entrepreneurs can create innovative models that ensure that there will continue to be new generations of skilled IT specialists with enhanced market culture that will either join the ranks of large IT companies or pursue new business projects.

What Did Not Matter (or Mattered Less)?

When synthesizing the challenges in the region that probably slowed down but did not stop entrepreneurship from rising, we must start with history and culture. Most of the members of the regional ecosystem interviewed during the writing of this book were convinced that societies in the Balkans do not provide a favourable setting for entrepreneurs. Part of that can be traced back to Socialism when self-initiative and entrepreneurship were considered as forms of going against the social interest. That deeply marked the thinking of older generations, who express suspicion and concern when younger family members decide to pursue entrepreneurship.

Despite the fact that the socio-cultural setting is not as favourable as elsewhere, this has not blocked dynamics and success of entrepreneurial activities. In that sense, initiatives in building awareness that promote the values of entrepreneurship and celebrate local heroes can make a difference. Culture and history matter, but as the experience of the Balkan countries shows alone, they cannot prevent entrepreneurial communities from thriving.

It is important to note that media has a big role to play here. As discussed earlier in this book, traditional media has not always been an immediate and active ally in entrepreneurship dynamics across countries in the region. This is one of the reasons that the change of attitude towards entrepreneurship has been so slow. This is also why many of the region's success stories are known mostly within the ecosystem and are not celebrated by societies at large. While a more active national media would have made things easier and faster, it is worth noting that entrepreneurial thinking has produced compensatory solutions. Entrepreneurial journalists and influencers have used digital platforms and created virtual media outlets that have become the voice of entrepreneurship communities across the region. Even without the reach of traditional media, they have managed to connect with audiences outside of the ecosystem and stimulate entrepreneurial curiosity among young people especially.

Another big topic for the region is the role of academic and research institutions. On one side, one of the fundamental ingredients for the success of the regional entrepreneurship is the tradition in good education in STEM domains. In that sense, good education is a necessity when it comes to growing an entrepreneurial ecosystem. On the other side, educational reforms take time. Ministries and academic institutions, which need to initiate and lead such reforms, tend to be conservative and slow when it comes to change. Our observation is that once there is a vibrant community with active entrepreneurial culture, it finds ways to compensate for the slowness of the educational sector. Through the initiation of entrepreneurial projects that create alternatives to formal education programmes in traditional universities and institutes, the ecosystem has started to train more and more students with relevant skills in areas such as entrepreneurship, innovation, data analytics, sciences, robotics and so on.

Another theme that was a favourite for pessimistic voices in the region is the small size of markets of Balkan countries that represent a limiting factor for new ventures. All members of the entrepreneurial communities interviewed for this book were asked to analyse the consequences of Balkan markets as these are comparatively small and distant from developed entrepreneurship centres. While some argue that this is an objective challenge for the development of entrepreneurship in the region, others suggest that every challenge must be treated as an opportunity and that local markets come with the advantages of location at the crossroad between Europe and Asia, as well as the opportunity to pursue higher level of distinctiveness and specialization, which minimizes the importance of market size. Generally, scale and location per se are not perceived as significant deterrents for developing vibrant entrepreneurial communities so other regions with limited market size should not take that as an excuse for not supporting local entrepreneurial efforts. Figure 13.2 contains the list of factors most frequently mentioned as factors facilitating or suppressing the growth of entrepreneurship in the Balkans.

Entrepreneurship as Spirit and Way of Thinking

One of the most important conceptual discussions that we have had about what is entrepreneuring and what it means to encourage an entrepreneurial mindset is whether it should lead to new venture creation only or to a different approach of solving all sorts of problems. We subscribe to the latter view and believe that for young ecosystems, especially such that are rising in less market-proficient socio-economic contexts, this view is essential. We believe that this framing is much more relevant for three main reasons. First, a new venture creation is typically the outcome of a new way of filling an unmet need, and it is the result of approaching the challenge or need with an entrepreneurial mindset. In that sense, nurturing entrepreneurial mindset in a society can expectedly lead to more startup projects but can also produce more entrepreneurial initiatives in all spheres of life such as existing or larger companies, civil society and government that will hire more employees with new ways of thinking and approach to problems.

Second, societies that are undergoing transitions similar to the ones experienced by the countries on the Balkans have less sophisticated market systems and associated support institutions than countries with more mature market economies. That may be treated as a challenge, but it also represents great opportunities. For instance, there are many social and market needs that are filled in countries with more sophisticated systems

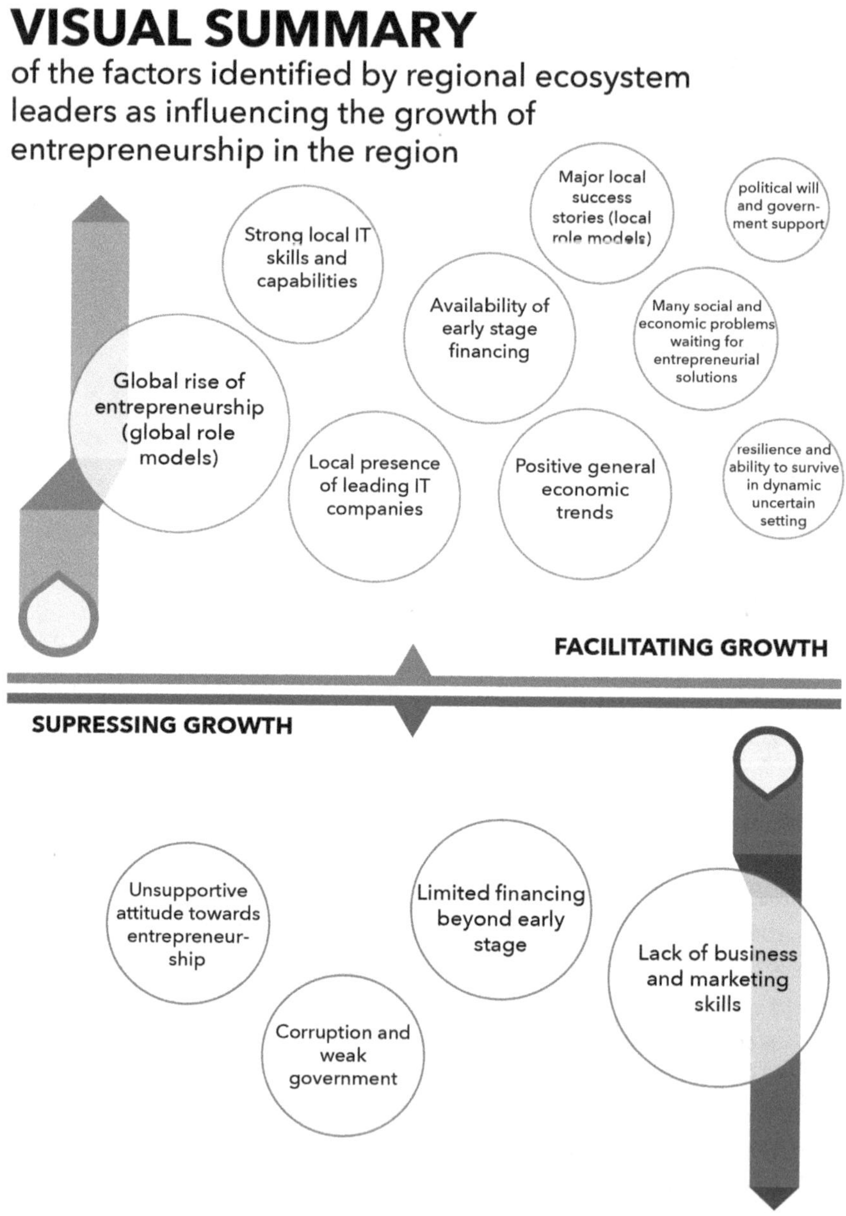

Fig. 13.2 Factors identified by regional ecosystem leaders as influencing entrepreneurship in the region

but remain unmet here. That creates a wide arena for applying entrepreneurial thinking and pursuing innovative solutions that go beyond private benefits and offer potential for public and societal value creation. This might be one reason why the pursuit of societal impact along with individual ambitions has caught up so fast in this region despite the short history of entrepreneurship.

Third, we live in times of disruption. Even advanced and developed societies with market economies are experiencing crises with existing social, economic and political systems that are prompting more and more calls for transformation. More and more citizens are questioning existing political systems and demanding fundamental change in the ways government leaders are elected and politics conducted. The rise of new generations with different priorities and fast-changing needs for skills in a world with rising robotization are causing frustration with educational systems and their resistance to change. The grand failures of the financial systems and inability to lower the share of the global population without access to its services have triggered numerous calls for redesign of this sector as well. In such dynamic times, a region like the Balkans needs more entrepreneurial minds that can help adapt and redesign systems on the go by addressing these contemporary problems. This is an alternative to the path of copying the existing systems of the leading economies only to later reach the same problems that they are experiencing today. In order to leapfrog in this global dynamic, the region needs more active entrepreneurs and a widespread entrepreneurial mindset.

Ambitions and Being Yourself

An important insight for us as authors of this book celebrating the rise of young entrepreneurial ecosystems is that every ecosystem has its own character that can only be fully understood and appreciated in the context of its own historical and societal dynamics. In this sense, while noting that the benchmarks against more mature ecosystems can lead to the formulation of important goals or occasionally serve as a source of inspiration, a simple comparison is often counterproductive. Young ecosystem should be geared as much on the outcome of early venture creation as on the process of shaping entrepreneurial mindset. In that, young ecosystems also have different goals, among which growing a community of entrepreneurs, policy-makers and supporters who are eager to experiment and embrace change with a healthy dose of ambition. In that sense, a higher rate of unsuccessful startups or write-offs may sound like negative news but can in fact be an indication of an important educational and growth process, which in our opinion is essential for success.

Growing an entrepreneurial ecosystem in a less sophisticated market context and bumpy political setting produces different compensatory and adaptation mechanisms. For example, while generally the lack of linkages between educational institutions and the entrepreneurial ecosystem is a hindrance, in the Balkans it has produced some exciting entrepreneurial projects. The rise of many vibrant academies, camps and summer schools for students and young professionals of all ages has gained popularity and has become an alternative to traditional education centres giving rise to experimentation and innovation in business models and curricula. In this sense, what seems to be a weakness of the ecosystem can also be seen as an opportunity, an arena for entrepreneurial experimentation that eventually becomes a new form of adding value to the entire ecosystem.

For these reasons, we hope that this book gives a rich and fair perspective of the transformative potential of entrepreneurship on the Balkans and celebrates the success stories of entrepreneurs in the regional context that has not been particularly supportive. Much has changed and the process has reached the tipping point where the transformation has to be acknowledged and widely celebrated. It is time to spread the good news and to celebrate local success stories and the heroes behind them.

Index

A
Academic and research institutions, 6
Accelerators, 11
Accelerator venture funds, 3, 35
Access to equity financing, 116
Acquisition, 18
Albania, 11
Altruism, 163
Amazon, 50
Angel investors, 12
App, 18
Apple, 50
Artificial intelligence, 49
Augmented reality, 49
Australia, 41
Automotive components, 51
Awards, 19

B
B2B ventures, 40
Balkans in world history, 8
Belgrade, 49
Berlin, 17
Beverages, 63
Bitdefender, 91
Bosnia, 11
Brain drain, 42
Branding, 40
Bucharest, 49
Budapest, 48
Bulgaria, 3, 11
Bulgarian Enterprise Growth and Investment Project (EGIP), 67
Bureaucracy, 24, 75
Business

V. Andonova et al., *Entrepreneurial Ecosystems in Unexpected Places*,
https://doi.org/10.1007/978-3-319-98219-9

incubators, 25
services, 51
training, 40
Business model, 118
design, 147
innovation, 137

C

Canada, 41
Capital asset pricing model (CAPM), 141
Car components sectors, 41
Cartelization, 43
Central and Eastern Europe (CEE), 17
Change, 145
Chinese, 18
Coding, 81
Cognitive schema, 158
Cold War, 8
Comecon, 41
Communication equipment, 51
Communist, 8
Communities, 122
building, 14
values, 9
Competitions, 85
Competitiveness, 42
Computer design, 81
Conflicts, 13
Consumer goods, 62
Copyright, 109
Corruption, 13
Council on Mutual Economic Assistance (CMEA), 62
Co-working spaces, 88
Creativity, 10
Croatia, 11
Cultural change, 14, 166
Cultural environment, 5
Curricula, 80
Customer acquisition, 119
Cybersecurity, 40

D

Deep tech, 50
Diaspora community, 44
Digital economy, 41
Digital Economy and Society Index (DESI), 28
Digital skills, 28
Dividends, 39
Donor programmes, 65

E

Early adopters, 163
Early-stage entrepreneurs, 14
EBRD transition index, 43
Economic downturn, 82
Economic performance, 12
Ecosystem, 8
Educational effect, 12
Education policy, 61
Electric vehicle technology, 102
Emerging market, 40
Emigrate, 48
Emory University, 11
Employment growth, 48
Energy solutions, 64
Engineering, 38
Enhancv, 158
Entertainment, 63
Entrepreneurial ecosystem, 5
Entrepreneurial finance, 6
Entrepreneurial mindset, 6
Entrepreneuring, 139

Entrepreneurship Database Program, 11
Entrepreneurship education, 28
Entry cost, 38
Environment, 9
Equity, 68
  and debt instruments, 34
  financing, 115
Escreo, 158
European Council, 8
European Investment Fund, 34
European Union, 8, 41
Executive education, 81
Executive hire, 130
Exits, 6
Experian, 85

F

Failed startups, 81
Fall of the Berlin Wall, 61
Farmhopping, 156
Fast-moving consumer goods, 63
Fear of failure rate, 49
Fintech, 30
Food, 63
Foreign investors, 43
Forging identity process, 160
Former Yugoslavia, 61
For-profit organizations, 106
Founders, 115
Frameworks, 101
France, 48
Free market economy, 43
Fund instrument, 11

G

Game companies, 18
Geopolitical challenges, 103
Germany, 41, 48
Gjirafa, 83
Global companies, 39
Global Competitiveness Report, 42
Global Entrepreneurship Monitor survey, 140
Global Innovation Index, 27
Globalization, 95
Global markets, 14
Google, 50
Governance, 42
Government
  agencies, 6, 74
  programmes, 6, 26
Great Britain, 41
Greece, 11, 60
Growth Cycle, 72–73
Guarantees, 68

H

Hackathons, 85
Hardware, 50
Headquarters, 106
Healthcare, 51
Helsinki, 48
Herzegovina, 11
Historical events, 10
History, 50
HP, 68
Human capital, 28
  advantage, 90
Human Rights Watch, 45

I

Immigration, 40
Industrial clusters, 46
Infrastructure, 6
Innovation, 29

Intellectual elites, 46
Internal academies, 79
Internationalization, 40
Invention-based ventures, 114
Investment
  bank, 18
  funds, 4
  philosophy, 150
  strategies, 142
Investors, 4
Ireland, 48
Israel, 47

J

Joint European Resources for Micro to Medium Enterprises (JEREMIE), 11
Journalists, 45
Junior Achievement, 80

K

Key entrepreneurship indicators, 20
Knowledge transfer, 47
Kosovo, 11, 103

L

Labour
  costs, 32
  market, 28
LAUNCHub, 11
Learning, 118
Legal and commercial context, 88
Legal status, 141
Lifecycle model, 19, 117
Living standards, 42
Loans, 68
Location, 5
London, 48

M

Macedonia, 11
Market
  dynamics, 23
  openness, 23
  size, 85–87
  stakeholders, 148
Marketing, 81
Master of Business Administration (MBA), 81
mBrainTrain, 159
Media, 14
Meetups, 104
Melissa Climate, 161
Mentors, 13
Mergers and acquisitions, 18
Metaphor, 8
Micro, 57
Microsoft, 66
Monetization, 163
Motivations, 157
Multinational corporations, 51
Musk, Elon, 64

N

National Experts Survey (NES), 21
Networking, 14, 78
Neveq, 34
New ventures, 10
Next generation, 81
NGO sector, 13
Non-profit, 116

O

Online payment platform, 66
Organizational forms, 141
Organizational identity, 146
Organizational sociology, 171
Outsourcing, 51
  software, 40
Overcapacity, 119

P

Paris, 48
Patent, 109
Personal values, 145
Phenomenological approach, 153
Philanthropic, 110
Pivots, 119
Playground Energy, 156
Policy initiatives, 76
Policy-makers, 4
Post-Communist governments, 43
Practitioner approach, 6
Prague, 48
Price sensitivity, 52
Private
  capital, 35
  fund managers, 75
  property, 9
  sector, 6
Problem solving, 10, 119
Productivity, 42
Product–market fit, 6, 119
Profits, 39
Publicity, 64

R

Raw materials, 62
Regional institutions, 14
Regionalization, 70
Relationships, 10
Represent a laboratory, 9
Reputation, 123
Retailing, 51
Returnees, 44
Returns, 17
Revenue model, 153
Rimac Automobili, 102
Robotics, 81
Role models, 14
Rumania, 11

S

SAP, 77
Scaling, 6, 119
Scenario planning, 10
Science parks, 25
Scientific approach, 6
Second World War, 9
Seed, 11, 34
Seed financing, 70
Self-employment, 57
Sensemaking, 12
Serbia, 11
Serial entrepreneurs, 68
Shared value, 138
Silicon Valley, 13
Skype, 84
Slovenia, 11
Small enterprises, 57
SMEs, 34
Social and cultural norms, 23
Social and/or environmental impacts, 114
Social capital, 123
Social entrepreneurship, 138
Social media, 45

Societal wellbeing, 8
Socio-cultural setting, 93
Sofia, 17
Software, 18
Soviet bloc, 41
Specialization, 62
Stakeholders, 5
Startup, 3
Startup Genome Report, 117–120
Startup hubs, 70
Startup landscape, 71
State ownership, 50
State-sponsored entrepreneurship programmes, 25
STEM-focused education, 41
Stigma of failure, 94
Stockholm, 48
Stock trading, 31
Stories of success, 45
Strategic thinking, 10
Strawberry Energy, 158
Structural problems, 43
Success stories, 13, 101
Swisscontact, 82
Systemic perspective, 5
Systems perspective, 134

T

Talent pool, 26
Taxes, 24
Tax rates, 39
Team building, 119
Team Formation, 160–162
Tech hubs, 49
Technological entrepreneurship, 13
Technology-driven, 12
Tech sector, 17
Tel Aviv, 93
Tickey, 83
Trademark, 109
Transfer capabilities, 6
Transition to market economy, 62
Trust, 41

U

Unemployment, 42
United Kingdom (UK), 48
United States Agency for International Development (USAID), 67
United States of America (USA), 41
Universities, 14, 29

V

Value proposition, 118
Venture capital, 31, 34
Venture creation intent, 157
Vilnius, 49
Viral growth, 130
Virtual reality, 49
Virtuous circle, 10
VMware, 85

W

Walltopia, 77
Warsaw, 48
Wellbeing, 140
Women as founders, 116
Workforce, 70
World Bank's Ease of Doing Business Ranking, 43

FSC
www.fsc.org
MIX
Papier | Fördert
gute Waldnutzung
FSC® C083411

Kolibri 360 GmbH
Ferdinand-Jühlke-Straße 7
99095 Erfurt, Deutschland
produktsicherheit@kolibri360.de